STORM DRAINAGE

STORM DRAINAGE

An engineering guide to the low-cost evaluation of system performance

PETE KOLSKY

Practical Action Publishing Ltd
27a Albert Street, Rugby, CV21 2SG, Warwickshire, UK
www.practicalactionpublishing.org

First published 1998\Digitised 2008

ISBN 13 Paperback: 9781853394324
ISBN Library Ebook: 9781780446059
Book DOI: http://dx.doi.org/10.3362/9781780446059

A catalogue record for this book is available from the British Library.

Since 1974, Practical Action Publishing has published and disseminated books and information in support of international development work throughout the world. Practical Action Publishing is a trading name of Practical Action Publishing Ltd (Company Reg. No. 1159018), the wholly owned publishing company of Practical Action. Practical Action Publishing trades only in support of its parent charity objectives and any profits are covenanted back to Practical Action (Charity Reg. No. 247257, Group VAT Registration No. 880 9924 76).

Typeset by Dorwyn Ltd, Rowlands Castle, Hants, UK

Contents

Acknowledgements

This manual is the main product of Research Project R 5477, 'Performance-Based Evaluation of Surface Water Drainage in Low Income Communities', funded by the Engineering Division of the British Department for International Development (DFID). DFID, however, can accept no responsibility for any information provided or views expressed. I would particularly like to thank David Crapper, Mike Slingsby and especially Brian Baxendale of DFID's Urban Poverty Office in New Delhi, who gave support to the project at varying stages.

I am very grateful for the collaboration of my colleagues on this research project. I owe a particular debt to Dr David Butler and Mr Jon Parkinson of the Imperial College of Science, Technology and Medicine who have worked with me on the collection and analysis of data, and in thinking through practical issues in our field work. They have been my closest collaborators. Mr George Heywood, now of TyneMarch Systems Engineering Ltd, performed some of the earliest useful analyses of field data, and his findings are used extensively in Chapter 3. I also owe a great debt to the support and stimulation of my colleagues in the Environmental Health Group of the London School of Hygiene and Tropical Medicine, particularly Dr Sandy Cairncross for sound support of all kinds throughout the project, and to Dr Ursula Blumenthal for critical support and help in thinking through the most difficult stage of the work. A number of individuals were kind enough to read drafts of this manual and all offered useful suggestions for its improvement. These included Brian Baxendale, David Butler, Andy Cotton, Sandy Cairncross, Mark Harvey, Jon Parkinson and Kevin Tayler. I must also acknowledge the support of varying kinds from Wallingford Software; the firm has been generous in allowing use without charge of its high quality drainage simulation software WALLRUS, SPIDA and HydroWorks, and its director, Dr Roland Price, has offered technical and moral support throughout the project.

I owe a great debt to my collaborator Professor T.A. Sihorwala of the Shri Govindram Seksaria Institute of Technology and Science, as well as to the Institute's director for support in establishing a local research team, and creating support for our research in Indore. The local team constituted the Indore Drainage Evaluation Project, and worked hard in the often chaotic and unpleasant conditions of monsoon floods; without their work there would be no manual. I therefore wish to acknowledge individually the contributions of engineers Mansur Ali, Prakash Chaudhari, Hakimuddin Haidery, Ravi Jain,

http://dx.doi.org/10.3362/9781780446059.000

Dilip Sharma, Ajay Shukla and Shailendra Tiwari who helped me to try out the ideas described in this manual; Jagdish Gharu, and Shyamlal and Rajesh Shiriswal also provided essential field support without which the work would not have been done. Some of the most interesting lessons came from the social sciences, the result of intense field work and analysis by Rajesh Pathnaik, Carolyn Stephens and Simon Lewin in their studies of risk perceptions of flooding in low income communities. Himanshu Parikh, the consulting engineer on the Indore Slum Improvement Programme, has been generous in sharing both information and innovative thinking on drainage design; the research conclusions broadly support the wisdom of his foresight and approach.

The field work would have been impossible without the patient support, help, and contribution of many in both the Indore Municipal Corporation and the Indore Development Authority. Within the Indore Development Authority, I would particularly like to acknowledge the support of the director, Mr Dagao Nkar; executive engineer Dillip Agashe, and engineers Uday Mane-Patil and Chandra Pamecha. Within the Municipal Corporation, I am especially grateful for the help of engineer Tripathi in the early stages of the work, and of Dr Puranik for insight into both the public health and solid waste problems of Indore.

All of the above have contributed time and thought to this work; errors and omissions, however, are the sole responsibility of the author.

1 Introduction

Drainage and surface water drainage

DRAINAGE IS THE removal of unwanted water. There are many forms of unwanted water in a city, and therefore a variety of drainage requirements. (See Table 1.1.) *Runoff*, also called *stormwater*, is the fraction of rainfall that runs off the surface of the ground, and leads to flooding; surface water drainage removes runoff from the streets and homes of the city. Frequent small floods damage roads and housing, disrupt business, and represent a basic threat to human health. This manual is designed to help those working to improve surface water drainage in low-income communities in developing countries.

Table 1.1 Drainage requirements of a city

Type of Unwanted Water	*Characteristics and Problems*	*Physical Control Strategies*
Local runoff	high flow short duration, seasonal	drains, fill, storage, and pumping
External runoff (e.g. overflowing river)	high flow short duration, seasonal	dikes, fill, training walls
Ponds (from high groundwater)	long-term ponding	dikes, fill, pumping, drains
Sullage (domestic wastewater, exclusive of toilet waste)	low flow steady, not seasonal faecally contaminated via laundry, nappies, etc.	sullage ditches, discharge to storm drains, soakaways, sewers
Toilet wastes	low flow steady, not seasonal heavy faecal contamination	sullage separation and low-cost sanitation; septic tanks and soak-aways; sewers

Surface water drainage and public health

During a flood, runoff mixes with human wastes from sewers, drains, and latrines, and spreads them through the streets and homes of the community. These wastes carry the bacteria, viruses, and parasites responsible for a wide number of gastro-intestinal infections, including diarrhoea (which kills over three million children a year), typhoid, cholera, and intestinal worm infections.

All sanitation systems need good drainage; none can function effectively during a flood. Moraes (1996) studied the effectiveness of drainage and sewerage in the slums of Salvador, Brazil, and found that surface water drainage (which also carried some sewage) reduced the odds of frequent diarrhoea by more than 50 per cent; surface water drainage combined with low-cost sewerage reduced the rate even further.

Moist soil provides an ideal environment in which worm eggs, passed in the human wastes, can mature and spread to new victims. Poorly drained and unpaved sites may thus promote the spread of a number of worm infections, such as roundworm, hookworm, and whipworm. These infections can lead to anaemia, stunting, and poor nutrition, as the worms grow at the expense of the infected child. Smith (1993) studied the health impact of sewerage in the Palestinian West Bank, and concluded that an otherwise unlikely *increase* in roundworm infection may have been caused by frequent flooding of the sewer during storms, due to inadequate surface water drainage. In addition to reducing diarrhoeal disease, Moraes (1996) found that improved drainage reduced the prevalence of roundworm and whipworm by a factor of two, and hookworm by a factor of three.

The safety of drinking water can also be affected by drainage. Many water supplies in developing countries are intermittent, that is, there are times when the water in the pipes is not under pressure. If a flood occurs when the water is unpressurized, flood waters can enter the pipes through leaks. As flood waters are usually contaminated with human wastes, this can lead to severe pollution of the water supply. This is frequently cited as a cause of disease outbreak in Indian cities during the monsoon.

Ponding and blocked drains often provide breeding grounds for mosquitoes and habitat for snails. Poor drainage is therefore often associated with malaria and schistosomiasis, although the nature of these problems depends very much on local conditions.

The purpose of the manual

This manual was written to help engineers understand surface water drainage problems more clearly so that they can work on more realistic solutions. Many standard engineering texts discuss surface water drainage (e.g. Fair *et al.*, 1966; Escritt, 1972; Clark *et al.*, 1977; Khanna, 1992), and some (Bartlett, 1976; Cairncross and Ouano, 1991; WEF/ASCE, 1992) deal exclusively with the subject. Almost all look at drainage from the perspective of the industrialized world; only two (Watkins and Fiddes, 1984; Cairncross and Ouano, 1991) are written specifically for developing countries.

This book focuses on three questions of particular relevance to low-income, urban areas in developing countries:

- What is drainage performance? What happens when it floods?
- What are the effects of solids in drains upon performance?
- How can we evaluate a drainage system, to assess how best to improve its performance?

Who this manual is for

It is hoped that a variety of professionals will find this manual useful. First and foremost, it is aimed at *municipal engineers* who need to study and solve problems in their drainage systems. It should also be of interest to *consulting engineers*, who, while relieved of the responsibility for operation and maintenance, still need to think through questions of capacity and performance. Finally, the manual may be of use to *engineering instructors and students*, as a supplement to more traditional engineering texts on drainage.

The structure of the manual

The main purpose of this manual is to describe ways to evaluate drainage systems to improve performance. Before starting field work, however, it is essential to understand what sorts of things to look for and why. For this reason, the manual begins in Chapters 2 and 3 with a general review of the factors that affect flooding in urban catchments. Chapter 4 considers general approaches to drainage evaluation, in which Table 4.1 provides a 'roadmap' to the methods described in the remainder of the manual.

Chapter 5 presents methods for defining and studying the characteristics of the catchment area of a system. Chapter 6 examines ways to assess the nature of the flooding problem in a catchment. Chapter 7 outlines simple methods for estimating the magnitudes of flows that will pass through the drains, and includes an annexe describing approaches to the use of limited rainfall data. Chapter 8 presents ways to estimate the capacity of the network, and stresses the need to consider the *actual* state of the system, rather than simply its original design. Chapter 9 presents field methods for studying the network's structural condition, while Chapter 10 describes the evaluation of its operation and maintenance. Observation of drainage networks in action during storms is described in Chapter 11. Chapter 12 summarizes the main points and conclusions of the manual, as a guide to how to improve performance through evaluation of the drainage system. Table 1.2 is a short 'road-map' to the manual.

Origins of the manual

This manual is the outcome of two-and-a-half years of field work in the city of Indore, in Madhya Pradesh, India. The Indore Development Authority has undertaken extensive work on slum improvement, and has been supported in this work by the UK government. This has included both financial support from the British Department for International Development (DFID), and technical assistance from the Urban Poverty Alleviation Field Management Office of the DFID in Delhi. DFID staff were aware of the importance of drainage in urban upgrading schemes, and supported the idea of developing suitable evaluation techniques. Staff at the Indore Development Authority

Table 1.2 Structure of this manual

Stages of the Study	*Relevant Chapters*		
Background	1 Introduction	2 Drainage systems, flooding, and performance	3 Factors that affect performance
Overview	4 Drainage evaluation: general approaches		
Study the problem	5 Studying the catchment	6 Assessing flooding as a problem	7 Flow estimation
Study the solution	8 Assessing drainage capacity	9 Drainage network structural survey	10 Maintenance surveys
Observe performance	11 Studying drainage systems in action		
Conclusions	12 Summary and conclusions		

were also interested to assess the new ideas being tried in their projects, where roads were designed as a major part of the surface drainage system.

This project was primarily a collaboration between three research institutions:

- the London School of Hygiene and Tropical Medicine (UK)
- the Imperial College of Science, Technology, and Medicine (UK), and
- the Sri Govindram Seksaria Institute of Technology and Science, Indore, Madhya Pradesh, India.

As part of the research, engineering staff were recruited as members of the Indore Drainage Evaluation Project, which developed and tested the methods outlined in this manual over the two monsoons of 1993 and 1994. Given its history, it is not surprising that this book has a strong south Asian flavour. Nevertheless, most of the problems discussed are common in other parts of the world, and the lessons from the research are broadly applicable.

2 Drainage systems, flooding, and performance

ENGINEERS EVALUATE DRAINAGE systems to find out how well they work, and to learn how to make them work better. Before looking at detailed evaluation techniques, it is useful to think about how drainage systems actually behave during a flood. This chapter therefore begins with a review of what happens during a storm, and the linked processes of drainage and flooding. In many systems, events in dry periods *between* storms can affect a system's performance, so the chapter also looks at dry-weather processes that affect drainage capacity. The chapter then looks at community perceptions of flooding and 'performance', and how conventional engineering practice generally fails to answer their most basic questions. The chapter concludes with an approximate definition of drainage system performance used in the rest of the manual, and some practical limits in the use of performance measures and indicators for evaluation.

Wet weather processes: what happens when it rains?

Rainfall becomes runoff

Surface water drainage controls *runoff*, the portion of rainfall that runs off the surface of a drainage area and ends up in streets, drains, and natural watercourses. Drainage systems fail, and flooding takes place, when the *rate of runoff* or *discharge* (expressed in m^3/sec or litres/sec) exceeds the capacity of the drainage system. No two storms are exactly alike, and the rate of runoff can vary dramatically over the course of a single storm. The most significant physical factors of rainfall and runoff that affect drainage performance are considered below.

Rainfall is irregular and unpredictable. Studies of historical records can give some idea of the overall size of storms, but these cannot be used to predict the pattern of any particular storm or season. The most useful data are *rainfall intensities* (in mm/hr), as the rate of runoff or flow to be carried by the system varies with intensity. The surface area of the catchment can be seen as the upstream end of the drainage network; rainfall intensity would then correspond to the 'velocity' of flow (height of water/time) at the upstream end (Figure 2.1). Over a long period of time, a steady rainfall intensity would thus produce a steady rate of runoff, while increasing the intensity would increase the rate of runoff.

http://dx.doi.org/10.3362/9781780446059.001

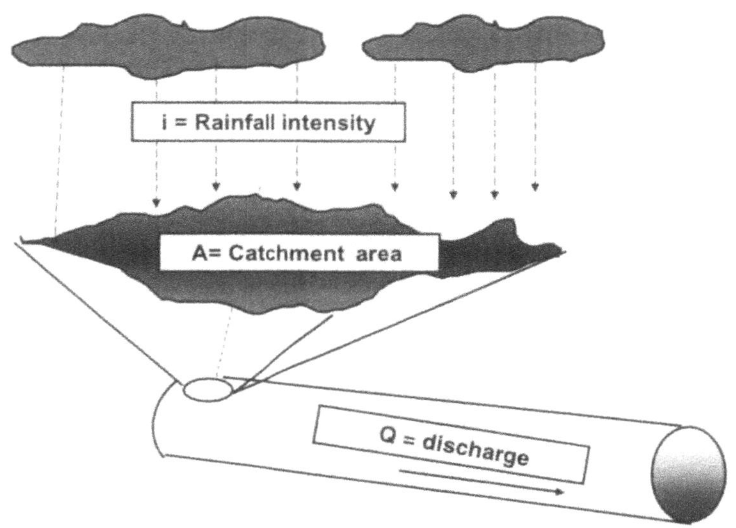

Figure 2.1 *Relation between drain discharge and rainfall intensity*

This intuitive view is borne out by the simple and widely used rainfall-runoff model of the rational (or Lloyd-Davies) method (for further details see Escritt, 1972; Bartlett, 1976; Chow *et al.*, 1987; Cairncross and Ouano, 1991). The rational method is based on the relation:

$$Q = 2.78 \, CiA$$

where

Q = the peak runoff discharge at a point (litres/sec),
2.78 = a dimensional term to convert between different units of length and time,
C = a runoff coefficient (dimensionless) varying between 0 and 1.0,
i = rainfall intensity (mm/hr), and
A = catchment area tributary to the point in question (hectares).

As a useful rule of thumb, 25 mm/hr falling on a single completely impervious hectare of land (C = 1.0) yields 70 litres/second of runoff discharge.

Hydrological research over the years has developed many more accurate models for predicting runoff from rainfall (e.g. SCS, 1975; DoE/NWC, 1983; Huber and Dickinson, 1988). These models take into account many more factors, but are much more complex than the rational method. While these improved models are useful for simulating large or complex catchments, the rational method illustrates the most significant factors, namely the catchment area, the runoff coefficient, and the rainfall intensity.

Catchment area The catchment area is simply the total surface area on which rain falls before draining to the point at which flow is being estimated.

This is the simplest term in the rational method, but still requires careful thought and field work. Many drainage systems have developed from natural watercourses, ditches, or old drains for which no catchment areas were defined. Working out the catchment area requires a good understanding of both surface topography and any changes (e.g. drainage channels) which modify natural drainage patterns.

Defining catchment boundaries is particularly difficult in flat areas, as small errors or uncertainties in topography can make a big difference in the boundaries. The boundaries can also shift during floods, as floodwaters can spill over from one catchment into a neighbouring one. Despite these occasional difficulties, *defining the catchment is the first step in any drainage evaluation or design*. All estimates of flow depend on this, and failure to check boundaries in the field can lead to gross errors in design, excessive flooding, and needless damage, nuisance, and cost.

The runoff coefficient The catchment area is made up of a number of different surface types (e.g. roofing, roads, and grassy plots) each of which contributes varying proportions of runoff from the same rainfall. The runoff coefficient for each surface represents the fraction of rainfall falling on it that turns into runoff. As shown in Chapter 7, an overall catchment runoff coefficient can easily be computed from the surface coefficients, and this composite C is what is used in the rational formula above. C can be taken as 1.0 for impermeable areas (e.g. paving, or the area covered by houses). For other areas, Table 2.1 shows appropriate values of C, and illustrates some of the following trends:

- Impermeable soils (e.g. clays) generate more runoff than permeable soils (e.g. sand)
- Steep slopes generate higher runoff flows than flat ones
- Ponds and storage (including that on flat roofs) reduce peak rates of runoff
- Leaf and grass cover reduce runoff through storage, and evapotranspiration.

In urban areas, the impermeable area is by far the most significant source of runoff. A number of urban drainage models simply ignore all unpaved area, not only because it contributes less volume, but also because the contribution is substantially delayed behind the peak from the impermeable areas. As cities grow, more areas become paved, vegetation is reduced, and upstream drainage improves; all of these add to greater downstream runoff. In assessing an existing system, engineers therefore need to consider future changes in the runoff coefficient of the catchment due to urban development.

Rainfall intensity Rainfall intensities in tropical climates are generally higher than those in temperate climates (Watkins and Fiddes, 1984; Shaw, 1994). Two events studied in Indore show the importance of rainfall intensity (mm/hr), rather than total volume of rainfall (mm) (Figure 2.2). The earlier

Table 2.1 Approximate runoff coefficients for areas not covered by buildings or paving

	Soil permeability			
Average Ground Slope	*very low (rock – clay)*	*low (clay – loam)*	*medium (sandy loam)*	*high (sand – gravel)*
I. Humid regions				
Flat: 0–1%	0.55	0.40	0.20	0.05
Gentle: 1–4%	0.75	0.55	0.35	0.20
Medium: 4–10%	0.85	0.65	0.45	0.30
Steep: > 10%	0.95	0.75	0.55	0.40
II. Semi-arid regions				
Flat: 0–1%	0.75	0.40	0.05	0.0
Gentle: 1–4%	0.85	0.55	0.20	0.0
Medium: 4–10%	0.95	0.70	0.30	0.0
Steep: > 10%	1.00	0.80	0.50	0.05

(From Watkins and Fiddes, 1984).

storm, shown on the left, did not cause flooding; the later storm, with less than one-third of the rainfall of the earlier one, nevertheless caused extensive short-term flooding. The maximum rainfall intensity of the second storm was three-and-a-half times greater than the first, producing a greater instantaneous discharge of runoff.

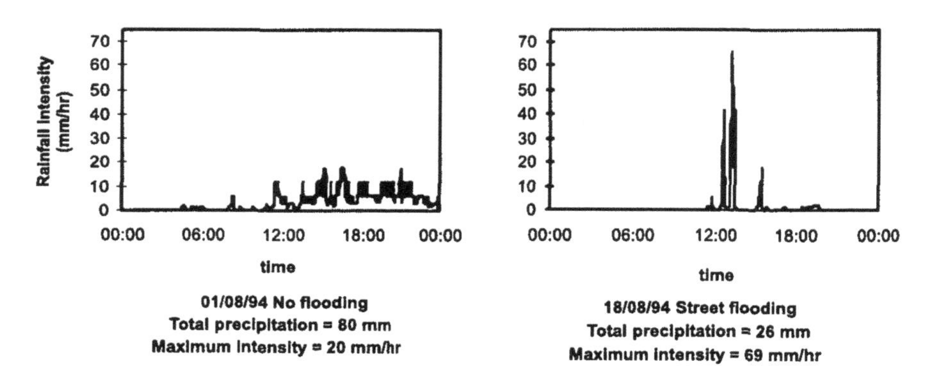

Figure 2.2 *Intensity and total rainfall in two events in Indore*

Intensity varies over the course of any storm; the rate of rainfall during the most intense ten minutes is naturally higher than the average rate of rainfall over an hour. In addition, intensity patterns vary from storm to storm, and high-intensity storms are less frequent than low-intensity ones. Intensity is therefore related to both duration and frequency, and systematic drainage evaluation is greatly helped by a set of *Intensity-Duration-Frequency (IDF)*

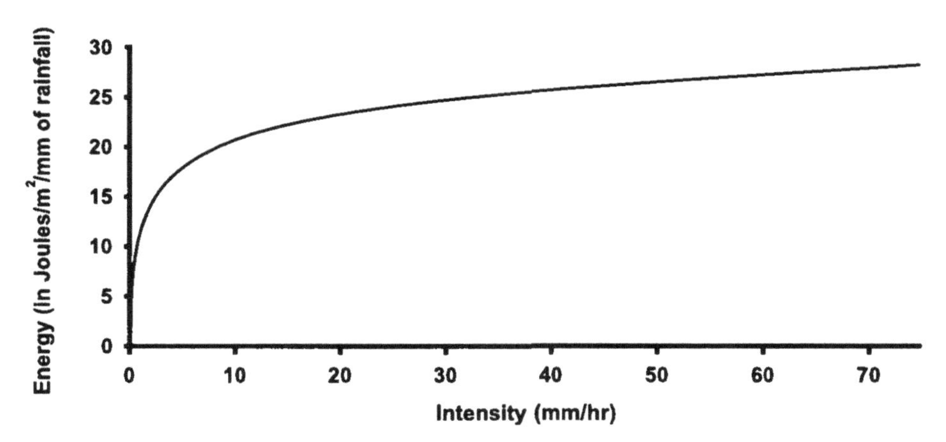

Figure 2.3 *Erosive energy vs. rainfall intensity (After Wischmeier, in Hudson, 1981)*

curves, when available. These curves show the maximum intensity of rainfall for a given duration and expected frequency. Unfortunately, rigorous preparation of such curves requires extensive data from recording rain gauges which are, in fact, rarely available. Some approximate guides for the development of IDF curves with limited data are presented in Annexe 7-A.

Rainfall intensity is also important in terms of 'erosive capacity', as shown in Figure 2.3. The rates at which soil is eroded per unit area is directly proportional to R, the 'erosivity' of the rain (USDA-SEA Agricultural Handbook 537, 1978, cited in Hudson, 1981). Erosivity depends greatly on rainfall intensity. Hudson stresses that erosion in tropical climates is more severe than in temperate ones because much more of the precipitation occurs at intensities exceeding 25 mm/hr; Hudson's method for computing erosion does not even consider rainfall intensities less than 25 mm/hr. While soil erosion may seem like a 'rural' problem, these findings have implications for unpaved streets and unstable soil in urban areas, where great damage can result from street erosion.

Runoff transports solids

During a storm, the impact of rain dislodges soil particles on the surfaces of the catchment, and runoff scours soil and picks up various solids as it flows towards larger drainage channels. Unstable slopes of soil often slip during rainfall, adding their load to the drainage system. As streams of runoff converge, the capacity to move solids increases. These solids are varied, and include soil, leaves and vegetation, rubbish, construction debris and road surfacing material. Some are picked up and carried in suspension, while others are dragged or pushed along the watercourse as bed load.

The effect of such solids on drainage performance depends on the quantity and nature of solids being transported, and on their fate during the storm. Such solids can:

- drop out before reaching a drain inlet
- drop out at the inlet, possibly blocking flow
- drop out in the drain, reducing its hydraulic capacity or
- continue as either suspended or bed load to the ultimate discharge.

There is a growing body of information about such solids (e.g. Binnie & Partners and Hydraulics Research, 1987; Verbanck, 1992; Verbanck, *et al.*, 1994) but the available data only look at the problem in urban catchments within the industrialized world. In low-income communities of developing countries, resources are scarce for solid waste management, paving, construction regulation, and urban erosion control, so the problem of loose solids on the urban catchment surface is correspondingly much greater. One of the main goals of the work in Indore was to explore the role of such solids in drainage performance and flooding.

Runoff enters the drain

Drains are designed to carry runoff away from the service area. To do this, inlets must first capture the runoff, and the capacity of a whole system can be affected by the design, location, and level of inlets. Experience in Indore and elsewhere suggests that poor location and design of inlets can be a major problem. In some cases inlets lie above the area they are meant to drain and so lead to ponding. In other cases, too many inlets allow too many solids into the network throughout the year, especially during the dry periods, which contribute to partial or complete blockage.

The rate at which the runoff is drained from the street is influenced by:

- the hydraulic capacity of the inlet
- the water level (hydraulic grade) in the drain beyond the inlet.

These in turn are influenced not only by the design and construction of the permanent structures, but also by their maintenance, and the flow patterns in the drain. Indeed, where the water level in the drain exceeds both the inlet crest and the outside water level, flood water flows *out* of the drain and into the street, and the 'inlet' becomes an 'outlet'.

Runoff flows through the drain

The flow and level of water through the drain is determined by the runoff pattern, the design and construction of the network, and the extent and type of solids blockage. The process is complicated, and powerful computer models have been developed to simulate the time-varying rainfall, runoff, drain discharge, and water level in the network (e.g. the Wallingford Procedure, described in DoE/NWC, 1983; SWMM, described in Huber and Dickinson, 1988). A simple way to see the effect of solids blockage is to compute the full-flow capacity of a circular channel under varying degrees of blockage (ignoring the greater roughness of the solids, which reduces capacity even further). This was done for a circular pipe using the Manning equation and

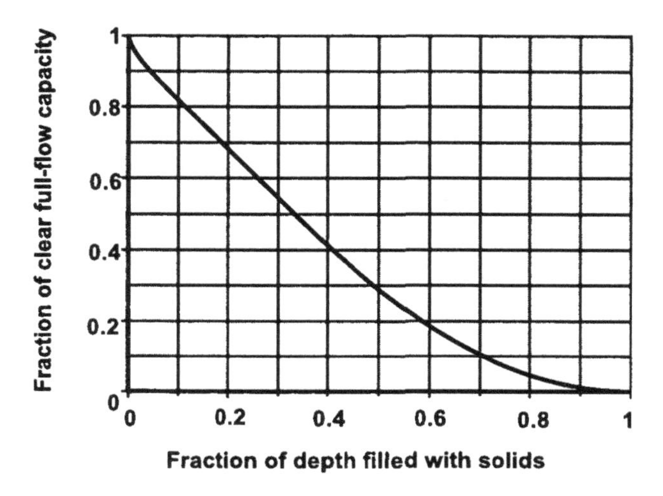

Figure 2.4 *Approximate pipe capacity reductions with increasing solids depth*

plotted in Figure 2.4. Note that under these simple assumptions, a 10 per cent blockage of the pipe results in nearly a 20 per cent reduction in capacity, while a 50 per cent blockage leads to a 70 per cent reduction in full-flow capacity.

Flooding takes place

Some flooding is inevitable in any drainage system during its service life; engineers cannot offer flood protection against every possible storm. As noted above, flooding can occur for a variety of reasons, including:

- high runoff exceeding design capacity
- limited downstream drainage capacity due to poor design, poor construction, poor maintenance or temporary blockages
- inundation from adjacent catchments with inadequate drainage capacity
- inadequate local inlet capacity, due to poor design, blockage, or poor inlet location/level
- high downstream water levels, due to high levels at the final discharge point, or heavy runoff from other catchments
- flooding from the drain itself from flows of other catchments backing up through the system.

The depth, area, duration and frequency of flooding can be used to define the performance of the drainage system, as explored on page 17. Performance reflects not only the interaction of the rainfall, catchment, and drainage network during the storm itself, but also dry weather processes (solid waste management, drain cleaning, etc.) which are considered in the next section.

When flooding takes place, runoff is no longer confined to the drainage network. Instead, runoff flows across the surface of the catchment along other paths which are rarely considered by the designer of the drainage system.

These paths, known as the 'major' drainage system have a very strong influence on the performance of the network during floods. The geometry of flow during a flood may be completely different from that of the drainage system designed by the engineer; the area of flow is much greater, and the direction of flow may be different (including through people's homes!). One of the main lessons from the research in Indore has been the need, during design and evaluation, to consider the effects of flows *above* the ground, as well as the performance of the 'proper' drainage network of pipes and channels designed by the engineer.

Flood waters are contaminated

Flood waters are heavily contaminated with human and animal wastes. Flooding of sewers, latrines, or areas where open defecation is practised spreads faecal contamination wherever the water flows. Water quality measurements of street flooding in Indore indicate up to 10^7 E. coli/100 ml in areas where sanitary sewers or combined drains were flooded, thus showing that the flood waters are as heavily contaminated as sewage. There was no significant difference in flood water quality between separate and combined systems during flooding; the contents of both are mixed with runoff and spill out on to the streets. Where flood waters rise above the building threshold, they can bring disease-causing organisms right into the home. As noted in the Introduction, the contamination of flood waters can also leak into the pipes of intermittent water supplies when the pipes are not under pressure.

Dry weather processes: solids deposition and drain maintenance

As noted above, some solids are washed into drainage systems during storms. Solids also enter during dry weather, and in Indore this appeared to account for the major portion of the solids deposited. This section outlines aspects of the deposition of solids in drains, and also some issues in their maintenance.

Solids deposition

Open drains Figure 2.5 shows solids deposition in an open drain in Indore over a one-year period from its initial cleaning during the monsoon of 1993. This graph shows the *average* build-up over 24 locations along the drain; different locations build up at different rates. The most important feature of the graph, however, is the gradual deposition over the dry season. In the dry period between the monsoons, solids were building up in the drain even when there was little or no rainfall. Several paired observations before and after storms also revealed no major changes in solids levels; large amounts of solids were *not* flushed out during the storms, or, if they were, they were replaced by others. These results therefore show that the major deposition of solids takes place during the dry period, as rubbish and construction debris find their way into the drain easily enough without being carried by runoff.

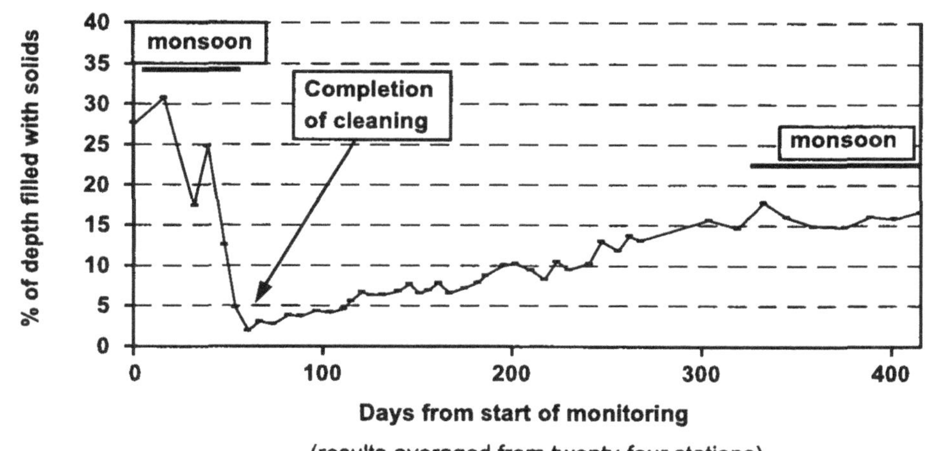

Figure 2.5 *Solids deposition over time in open drain in Indore*

Figure 2.6 shows the particle size distribution observed from six samples taken along the same open drain, and compares it with the distribution measured in the UK. The samples from the open drain in Indore are ten times coarser than those found in closed UK drains.

This difference can be explained by:

- open (Indore) and closed (UK) drains
- the different levels of solid waste management currently practised in the UK and India
- the greater extent of temporary road metal erosion, and construction debris in Indore.

While these results are from only a single open drain, drains in other low-income communities face similar conditions, and will encounter similar problems.

Closed drains Covering drains can reduce these problems, by restricting the entrance of solids along the length of the drain, and through the use of gully pots. They bring with them other problems, however, namely:

- restricted entry of runoff, where inlets are blocked
- more difficult cleaning
- more difficult identification of problems; it is much harder to see what is going on in closed drains.

Stormwater inlets to closed drains are essential, and these can be provided with or without gully pots (catch basins). Pots provide an area of lower velocity where some of the solids will settle out. Inlets without pots require less maintenance than those with them, but will not restrict the entry of solids

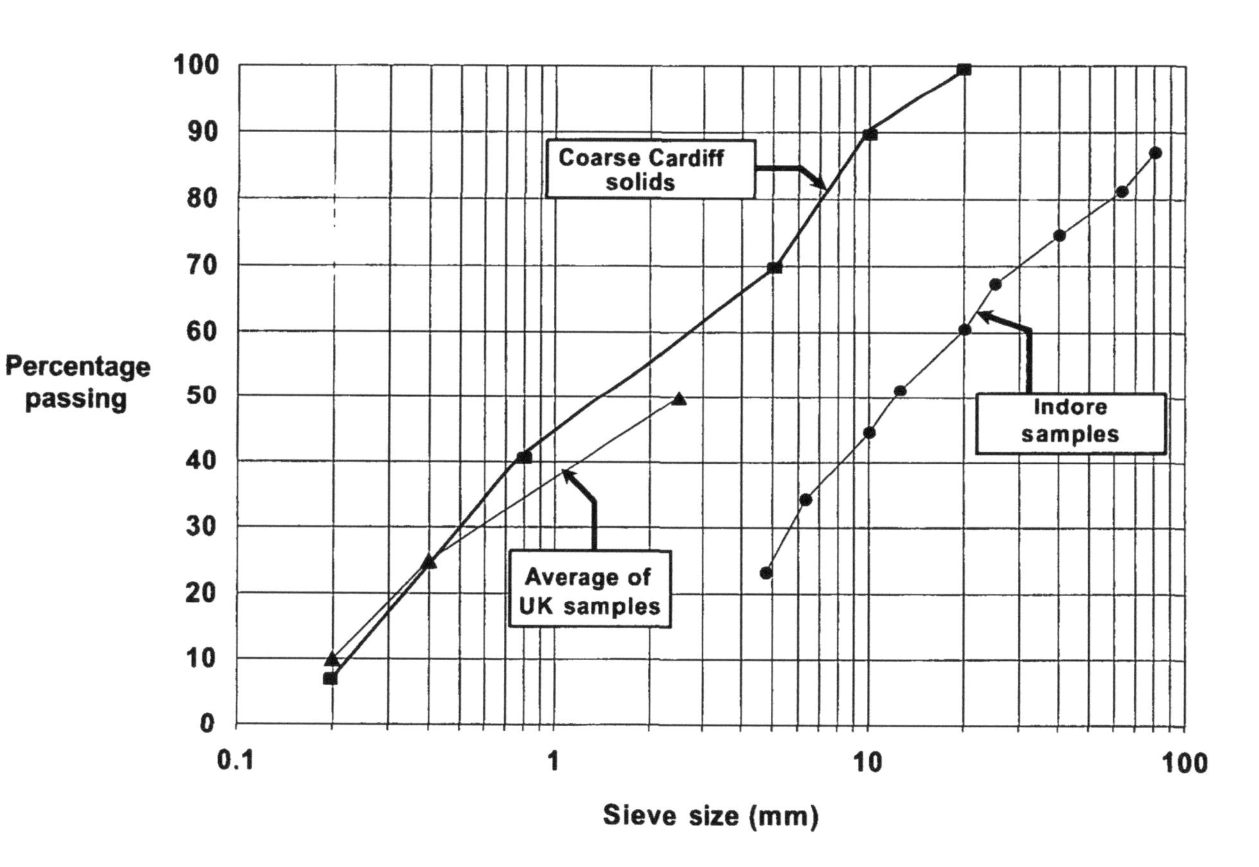

Figure 2.6 *Comparison of drain solids size distribution in Indore and the UK*

smaller than the inlet mouth or grating. Studies on UK 95 litre gully pots (Butler and Karunaratne, 1995) show that for equivalent intensities of about 25 mm/hr, over 90 per cent of solids greater than 500 μm are retained. Efficiency drops off rapidly for smaller sizes, and according to the theory developed, at higher flow rates (i.e. rainfall intensities). The depth of sediment trapped in the pot does not substantially affect performance until blockage occurs. The implication of this is that if pots are provided, regular maintenance is required to clean them out. To minimize excessive cleaning costs, this would be done ideally just before they are completely filled. If, however, inadequate maintenance is performed, there is little point in providing pots. As yet, there is no rational method for deciding the size of pot required in any particular situation.

Some designers place inlets at every manhole in a closed drainage system, as they think that 'every inlet helps', and the difference in cost between a closed manhole and an inlet may be small. Unfortunately, every runoff inlet is also an inlet for solids; rubbish, road metal, and construction debris enter drains at inlets, and the more inlets that are provided, the smaller the advantage of closed drains in keeping out solids.

Drain cleaning

Drains in many communities are often cleaned only in response to complaints. There is sometimes a special effort made to clear drains around the start of the rainy season. Given the high organizational cost of a drain-cleaning effort, it makes sense to concentrate the work in a single 'push' to clear the accumulated debris of the preceding year; cleaning halfway through the dry season would achieve less, as more debris would enter, necessitating the rainy season cleaning in any event.

Drain cleaning, like any other form of solid waste management, involves a number of stages:

- cleaning of drains, and piling the solids by the side of the drain
- collection and transport of solids piles
- disposal/reuse of solids at an appropriate site (e.g. landfill).

In practice, attention tends to diminish after the cleaning of the drain itself. Cleaning appears to solve the immediate problem of drain blockage, and eases political pressure from residents. In addition, cleaning is fairly simple; each individual or group of workers can clear a specific length of drain. By contrast, collection and disposal is far more complicated, as it requires organization of transport to visit a number of sites, and the handling of a wet and potentially dangerous waste.

Policies on drain cleaning vary, and practice often differs greatly from official policy. A common policy in Indore required that cleanings be picked up a week after their removal from the drain, so that the waste can dry out and be handled more easily. In practice, this separation of collection from removal often leads to piles drying by the side of drains for months before their removal. The resources of transport and organization required for the collection, transport, and safe disposal or reuse of drain cleanings are greater than those needed for the simple cleaning itself; the political or public demand for safe ultimate disposal, however, is far less than that for the cleaning.

Uncollected piles of drain cleanings create two problems:

Return to the drain Uncollected drain cleanings represent one of many forms of solid waste in streets which are likely to return to the drain during storms, or as a result of physical disturbance (from cars, lorries, children, roadwork, etc.) or by actions of residents fed up with the nuisance.

Public health Where such cleanings are from combined drains, they represent a serious public health hazard, as they contain a high concentration of disease-causing organisms which have settled out from the sewage. Children, the population most susceptible to gastro-intestinal infection, can often be seen playing with material from drain cleaning. After some external drying, drain cleanings appear fairly harmless, as they do not smell. Roundworm and other parasitic eggs, however, can be expected to survive several months in a pile of drain cleanings which, while dry on the outside, remains moist inside, providing an ideal environment for the development of the parasite.

Assessment and monitoring of drain cleaning practice can therefore be an important component of drainage evaluation.

Community perceptions of flooding

Slum residents who endure flooding see problems differently from technical staff who are responsible for drainage. Some of the differences stem from first-hand experience of flooding, while others come from the difference of technical and non-technical perspectives. While engineers may understand technical aspects of flooding better than residents, residents certainly understand better than anyone else how flooding affects their lives. Stephens *et al.* (1994) used qualitative social science research techniques to investigate the perceptions of flooding by members of four Indore communities. Some of their findings, summarized below, are of particular interest to engineers.

Priority

We need to keep flooding in perspective among the other difficulties of being poor. Even where flooding was frequent, it 'was ranked low in comparison to other risks and problems, such as improvements in job opportunities, provision of housing, mosquitoes and smelly backlanes'.

Predictability

> A major concern mentioned by residents of all four areas relates to the predictability of the flooding event . . . In other words, even extensive inundation is bearable if expected . . . Interventions aimed at ameliorat-

ing the effects of flooding should try to take account of these needs of the community to understand and adapt their coping strategies if necessary. [This reaction was echoed during the Easter 1998 floods in the UK, with the great dissatisfaction of victims over 'lack of warning'.]

Expectations

> [In areas of the slum improvement programme . . .] residents had high expectations of drainage improvement when the projects were initiated and appear to have expected that flooding and inundations would cease or be reduced substantially. It is likely that the feeling that their expectations have not been met is in part an immediate sense of disappointment that flooding has not ceased altogether.
>
> [Drainage] Interventions would gain favour if residents understood and were clearly informed about the effects (good and bad) on the environmental risk which they perceive as inherently a natural event. This, of course, necessitates technical personnel being able to predict the consequences of technical interventions. Such a strategy might reduce the scale of expectation.
> Stephens *et al*, 1994

These observations all come from within a single city during a heavy monsoon, and cannot be taken as 'rules of thumb' for elsewhere. Nevertheless, they all point to the need for dialogue between technical professionals and community members, to help technical staff understand what the community's problems are, and to help community members understand what drainage interventions can and cannot do about them. Most conventional technical approaches do not, unfortunately, include this dialogue, or help the engineer to answer the questions of residents.

Drainage performance and evaluation

From the point of view of the community, there are thus three main performance questions:

- How often does it flood?
- What happens when it does flood?
- When will it flood?

The third question of advanced warning is a critical one to residents, especially those living adjacent to major watercourses, where such warnings may be possible, and where residents may have already developed informal systems. Such systems deserve careful study, and municipal and hydrologic authorities need to consider whether these can be improved. This important subject, however, will not be further considered in this manual due to both its complexity and site-specific nature.

To answer the first two questions, the hydraulic performance can be defined in terms of:

- the frequency
- the depth
- the area or extent
- and the duration of flooding.

The work done in Indore used this definition as a starting point, and tried to evaluate the performance of several systems to see the effects of design and maintenance aspects upon performance. Some of these findings are presented in the next chapter. In everyday working practice, however, physical performance is both difficult to measure and difficult to interpret. Chapter 6 describes some of the methods used in Indore to obtain direct measurements of physical performance, but, with a few exceptions, these are too time- and energy-consuming to be manageable for ordinary system evaluation. In addition, analysis of such data can be complicated, and must be interpreted on an individual catchment basis; there are no general standards of drainage performance, because each catchment is unique. Other fields, such as public health and rural water supply, face similar problems in evaluating impact. An alternative for working evaluations is to examine *indicators* which, although not direct measures of the impact or outcome itself, can show approximately what this is likely to be. In everyday drainage practice, it makes sense to focus on process and performance indicators, because performance measurement itself is unwieldy.

This guide will explore:

- research findings on performance, particularly concerning solids,
- performance measurement for pilot study or research purposes, and
- other indicators to assess ways in which a drainage system can be improved.

Table 2.2 Types of performance measures and indicators

Type of Measurement	*Examples*	*Advantages*	*Disadvantages*
Performance	Depth Area Duration of flooding	Measures what you want to know	Difficult to measure Seasonal Link to action often unclear
Performance indicator	Solids levels As-built capacity Inlet blockage	Relatively easy to measure Clearer link to action than performance measurements	Relation to outcome may not be clear May measure symptom, not cause of problem
Process indicator	Frequency of cleaning Staff time committed to operation	Relatively easy and practical Can be routine Clear link to action	Relation to outcome may not be clear

3 Factors that affect performance

Before evaluating a drainage network, it is useful to think about which factors influence performance and how. This chapter therefore reviews several aspects of a drainage system, and their likely effects upon performance. These include:

- type of drainage system (major/minor, open, closed, separate, combined)
- hydraulic capacity (e.g. conduit size, shape, slope, extent of blockage)
- street grading
- inlet design (number, design, and location)
- catchment surface and storage (soakaways, roofs)

These factors are reviewed in light of existing literature, and our recent field research in Indore.

Types of drainage system

Major and minor drainage

As noted in the WEF/ASCE Manual (1992) on drainage:

> Every urban area has two drainage systems, whether or not they are actually planned for and designed. One is the *minor system*, which is designed to provide public convenience and to accommodate relatively moderate frequent flows. The other is the *major system*, which carries more water and operates when the rate or volume of runoff exceeds the capacity of the minor system. Both systems should be carefully considered.

Despite some important exceptions (e.g. Argue, 1986; Parikh, 1990; Tayler and Cotton, 1993; Watson-Hawksley and AIC, 1993), most drainage design focuses only on the minor system of engineered pipes and channels to prevent flooding. In almost all cases, the major system of roads and open space has not been designed or systematically considered, despite its critical role during flooding. Conventional designs focus on reducing the nuisance of relatively small and frequent storms by removing their runoff from the service area before significant flooding occurs. The weakness of this approach is that not enough attention is given to what happens when floods *do* occur.

Types of minor drainage systems

There are two overall characteristics of minor surface drainage systems that have a strong influence on maintenance and performance. These are:

http://dx.doi.org/10.3362/9781780446059.002

- their contents (only surface water, or surface water combined with sewage) and
- their construction (open or closed conduit).

In general, separate surface drains are easier to maintain than combined ones; they are dry most of the time and contain fewer faecally-contaminated solids. On the other hand, it often seems cheaper and easier to build and maintain a single combined drainage system, rather than two separate ones. This is particularly true where combined drains are already in use in other parts of the network, as there seems little benefit in separating flows upstream if they are only going to be recombined downstream!

Similar trade-offs are involved in comparing open and closed systems. Open drains are generally easier to build and clean; when problems do occur they are easier to identify, as the system is open to view. In addition, they do not have restricted entrances that block easily. Against this, open systems are much more exposed to the entry of rubbish, construction debris, and other solids. In addition, the slopes of open drains are severely limited by existing topography, making 'self-cleansing' velocities more difficult to obtain. Table 3.1 summarizes some of the trade-offs involved in comparing different drainage systems.

In summary, the type of system influences performance largely through maintenance issues of short-term blockages in storms, or long-term build-up

Table 3.1 Comparison of conventional drainage systems

Type	*When does it flow?*	*Safety from solids entry*	*Ease of cleaning*	*Risk of inlet blockage*	*Use of urban space*	*Problem with flat/ adverse ground slope*	*Impact of failure*
Separate Storm and Sanitary Drainage							
Open Storm	Storms	Poor	Best	Not a problem	Poor	High	Medium, Seasonal
Closed Storm	Storms	Good	Poor	Significant	Good	Lower	Medium, Seasonal
Open Sanitary	Daily	Poor	Medium	—	Poor	High	Severe, Constant
Closed Sanitary	Daily	Best	Poor	—	Good	Lower	Severe, Constant
Combined Storm and Sanitary Drainage							
Open	Daily	Poor	Medium	Not a problem	Poor	High	Severe, Constant
Closed	Daily	Good	Worst	Significant	Good	Lower	Severe, Constant

of solids through inadequate slope or difficulty of cleaning. The key question in assessing the suitability of any of these systems is 'how well can it be maintained?'

Hydraulic capacity

Hydraulic capacity is affected not only by the decisions of the design engineer, but also by the quality of both construction and maintenance. Slopes of drains after construction often differ from those specified by the designer, and uneven joints or debris left in place during construction can trigger a process of solids build-up which will later reduce performance. Similarly, poor maintenance leads to the build-up over time of solids, which can lead to more frequent flooding. Design errors can arise when the engineering study fails to consider the entire catchment area, or underestimates the runoff coefficient; under these conditions, flows are higher than expected, and the designed system is too small. While these are all different problems, they each affect performance through reduced or inadequate hydraulic capacity. This section outlines the effects of capacity on the frequency, depth, area and duration of flooding.

Frequency of flooding

Frequency is the most common design parameter for drainage systems. Common rules of thumb are that drainage system capacity in business districts should be exceeded only once in five years, while drains serving residential areas may be allowed to flood once every two or three years (Fair *et al.*, 1966; Cairncross and Ouano, 1991). Cairncross and Ouano point out that more frequent flooding than this may well be appropriate in low-income communities, if the alternative is no drainage improvement whatsoever. Tayler and Cotton (1993) agree, suggesting periods of one year or less for urban upgrading schemes.

It is generally assumed that frequency of peak runoff flows corresponds to the frequency of rainfall intensity; the five-year peak rainfall intensity of a given duration is assumed to produce the five-year peak runoff discharge for that duration. While there are good reasons for challenging this assumption (e.g. Sieker, 1979), it is sufficiently accurate to get a feel for the variation of frequency with capacity. Under this assumption, a graph of rainfall intensity vs. return period can be converted, via the rational method, to a curve of runoff vs. return period.

Figure 3.1 shows how rainfall intensity varies with return period in India, as reported by Kothyari and Garde (1992). They select 0.20 as the return-period exponent, but note that other researchers in other parts of the world have found values ranging from 0.18 to 0.26. Regardless of the exact value of this exponent, the curve shows a sharp increase in return period (and therefore a sharp decrease in frequency) for a small increment in rainfall intensity.

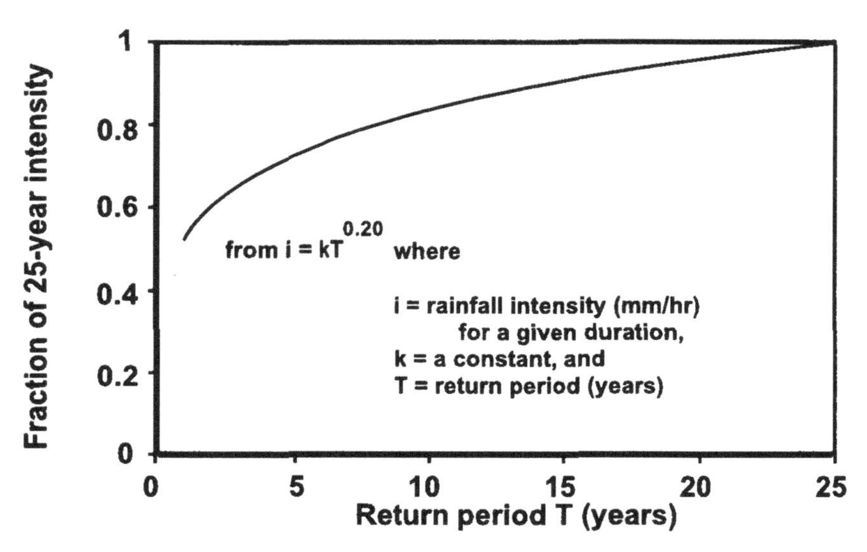

Figure 3.1 *Fraction of 25-year intensity as a function of return period*

If we accept the rational method for converting rainfall intensity to runoff peak flows, the graph and frequency-intensity relation can also be used to find:

- The relative frequencies of different peak flows
- The frequency of flooding of a drainage system of known capacity less than that of the 25-year storm. (For example, a drainage system with only 80 per cent of the capacity of the 25-year peak flow has a return period of about 12 years, and so will flood about twice as often as a system with capacity for the 25-year storm.)
- The relative flooding frequencies of different capacities of drainage systems.

The conclusions from this analysis (shown in Annexe 3-A) show that halving the frequency of flooding requires only a 15 per cent increase in drainage channel capacity. In the same way, a 13% *reduction* in capacity leads to a *doubling* of flood frequency, under the given assumptions. These results are summarized in Figure 3.2, which shows the increased frequency of flooding as capacity is reduced. The y-axis shows the factor by which flood frequency is increased by the percentage reduction in flow capacity on the x-axis; a 25 per cent reduction in hydraulic capacity thus increases the frequency of flooding nearly fivefold.

Figure 3.3 combines these results for capacity and frequency with those already shown in Figure 2.4, for the effect of varying solids depths upon hydraulic capacity. Figure 3.3 indicates that a circular drain designed to flood only once every twenty years will in fact flood ten times as often if solids take up 25 per cent of the pipe depth, and will flood on average twenty times as often if they take up 30 per cent. Even a modest 10 per cent depth of solids yields a doubling in flooding frequency.

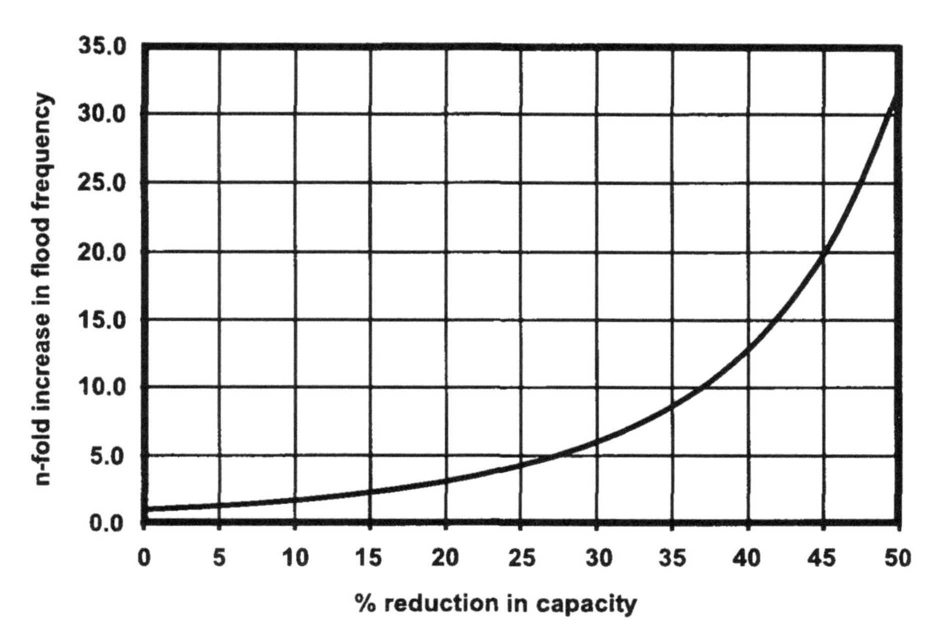

Figure 3.2 *Flood frequency increases as flow capacity is reduced*

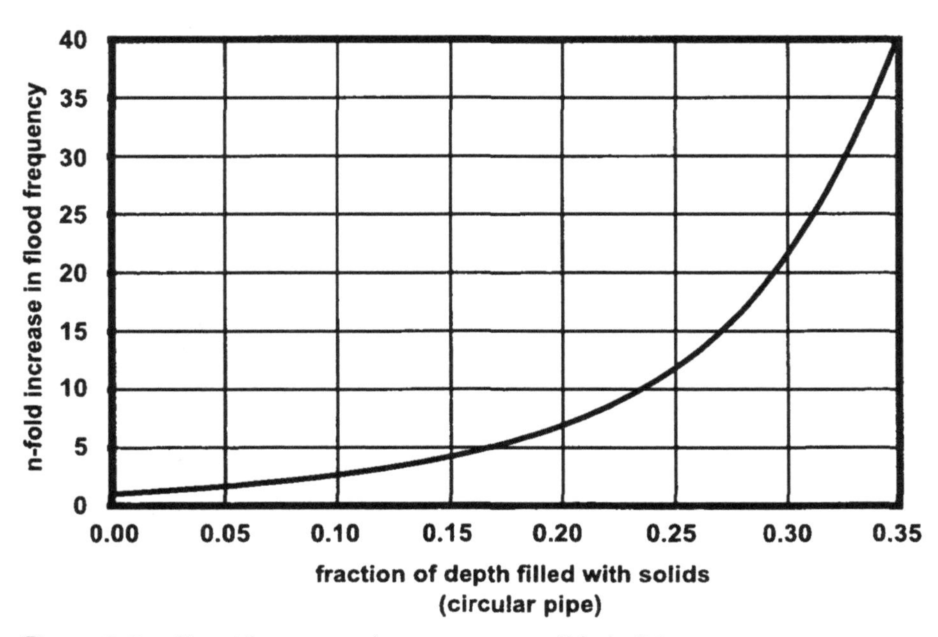

Figure 3.3 *Flood frequency increases as solids build up*

The hydrology used above is based on 'annual series', or the biggest storms of each year. This does not reflect the variation of rainfall intensity *within* a year, but only that *between* years. These two distributions will naturally differ, and are likely to vary from climate to climate. This matters where flooding is frequent, or the capacity of the system is reduced below that of the one-year storm. This can easily happen with, for example, a depth of solids of 15 per cent in a drain designed for a two-year storm. Figure 3.3 cannot, therefore, be used to infer the impact of solids on frequent flooding, but it suggests that solids may play a bigger role than most realize.

The preceding analysis is a theoretical one. Figure 3.4 is based on simulations of the performance of a conventional near-rectangular open-channel drainage system in Indore, as described in Kolsky *et al.* (1996). Rainfall events recorded over a seven-week period were used as input to a calibrated and verified hydraulic model, for which varying assumptions about levels of solids were made. Sample points (network nodes) were identified at uniform intervals along the drain, and used to indicate the extent of flooding. Figure 3.4 therefore shows the variation in flooding frequency and extent (as shown by the number of nodes flooded) at different levels of solids depth for the limited period of rainfall data collection *within* a single monsoon.

Figure 3.4 shows frequency of flooding to be much less sensitive to capacity and solids levels than in the analysis of Figure 3.3. The work in Indore shows, for example, only a two- to threefold increase in flooding frequency as a result of solids built up to 80 per cent depth during the study period.

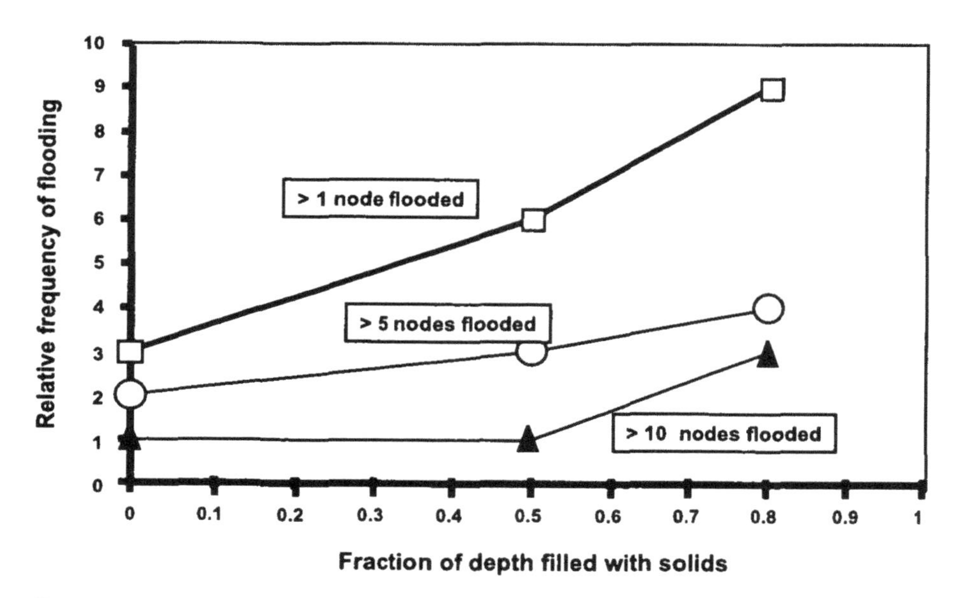

Figure 3.4 *Flood frequency vs. solids depth in an Indore catchment*

This results from a number of factors including:

- the inclusion of unsteady flow and storage effects in the hydraulic analysis of the Indore catchment, which are not present in the simple analysis of Figure 3.3
- the capacity of a rectangular open channel is less sensitive to solids levels than those of circular closed conduits
- the distributed nature of flooding along a drainage system, not reflected in the simple derivation of Figure 3.3, and
- the different statistical distribution of rainfall within a monsoon, relative to that of the largest events between monsoons; the equation $i = k(f)^{-0.20}$ is inapplicable when comparing events within years rather than between them. A review of the UK data of Folland and Colgate (1979) suggests that intensity varies more sharply with return period for more frequent events.

Figure 3.4 cannot be used for other catchments, or even other seasons in the same catchment, as it is unclear how much depends upon the particular catchment and season analysed. Nevertheless, the results show the need for better understanding of rainfall variation within years as well as between years for systems which flood often. Without such understanding, we are unable to predict the effect of improvements in capacity or maintenance upon frequent flooding.

Depth of flooding

Drainage capacity and solids levels generally have little influence on the depth of flooding *once flooding takes place*. This is because the depth of flooding is largely controlled by the much greater capacity and surface area of flow routes above the ground when flooding occurs. Figure 3.5 shows the limited effect of solids depth and (therefore hydraulic capacity) upon the depth of flooding in a catchment in Indore. Each of the lines in the graph shows the variation in flood depth at a different point along the main open channel drain; the situation thus differs across the catchment. In all cases, however, the difference in flood depths between a completely clean drain and one that was 80 per cent blocked was less than 10 cm.

Area of flooding

For a given urban catchment, the flooded area in any storm depends on the depth of flooding. This relationship is not always simple, however, particularly in flat areas. Large portions of the catchment may be at the same level, so that small changes in water depth (e.g. just overtopping the kerb level, or just overtopping the crown of a side street) can make large changes in flooded area. Once such critical levels are surpassed, there may be no further change in flooded area as depth increases until another critical level is passed. In steeper catchments, the flooded area may vary more continuously with depth of water. The quality of topographic data needed to predict this variation is rarely available. It may be useful for engineers to think through the effect of blockages or flooding in various parts of the catchment, and identify those

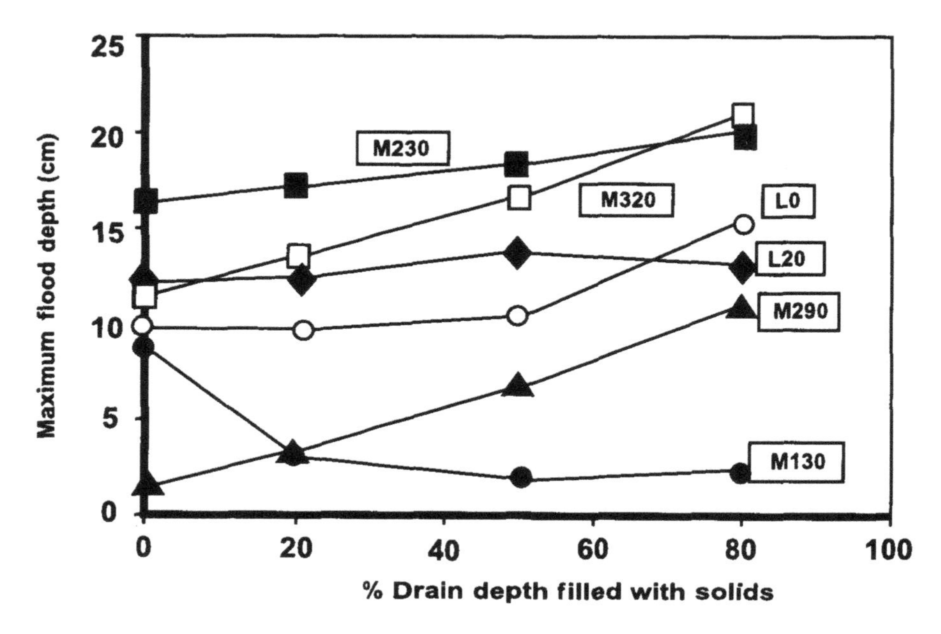

Figure 3.5 *Maximum flood depth vs. solids levels (Kolsky* et al., *1996)*

levels at which large (or important) increases in flooded area occur; when feasible, however, field observation during flood conditions can be far more instructive. An understanding of how flooded area varies with depth may lead to raising some plinth levels, or revising some street levels to reduce the vulnerability of certain areas.

The problem is further complicated by the distributed nature of flooding in a network; flooding often occurs in several places at once, and not just in one single pond. In our work in Indore (Kolsky *et al.*, 1996), we selected 15 evenly spaced points along an open drain, and simulated their flooding patterns during the monsoon under varying levels of solids in the drain. Figure 3.6 shows the number of these sample nodes which flooded under varying conditions of blockage, according to our simulations. This number does not necessarily reflect flooded surface *area*, as the exact variation between level and flooded area is not shown. Nevertheless, it is an alternative, more manageable, measure of 'extent' of flooding.

At all levels of solids, the graph is 'stepped'; in a clean drain, for example, one storm is big enough to flood ten nodes, while the next biggest storm is only large enough to flood seven. This reflects to some extent the irregular characteristics of urban flooded area vs. depth relations noted above. The distributed nature of flooding is also evident as some points are more vulnerable than others: with 80 per cent blockage one node is flooded nine times while four others were not flooded at all during the observation period.

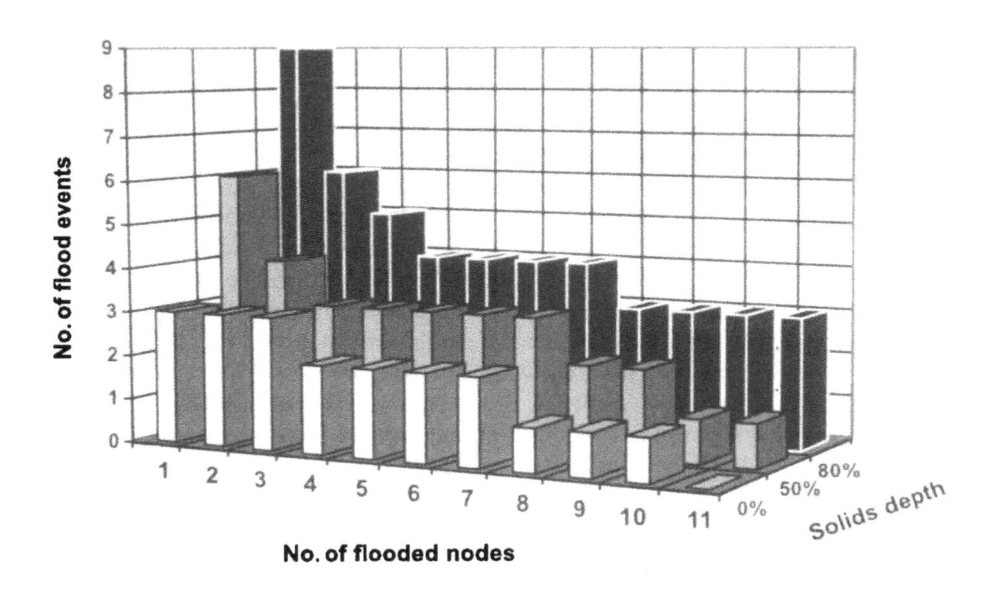

Figure 3.6 *Flood frequency and extent as a function of blockage*

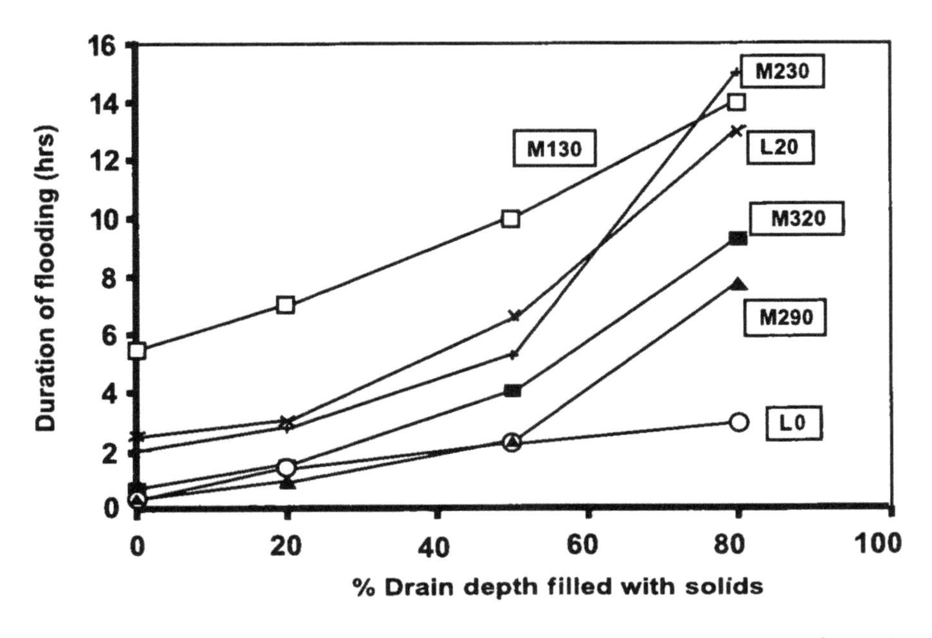

Figure 3.7 *Duration of flooding vs. solids levels in an Indore catchment (Kolsky* et al., *1996)*

Duration of flooding

How much difference do the pipes and channels of a drainage system make to the duration of a flood? The answer depends largely on surface topography. Where streets have good slopes, large cross-sections, and natural drainage routes, the capacity of the minor drainage system has little effect upon flood duration. On the other hand, in flat catchments where surface drainage is poor and only the minor drainage system can carry the flow, the duration of a flood depends greatly on the capacity of this system. Figure 3.7 shows the duration of flooding at selected points of an Indore catchment, and the effects of blockage (and implicitly of capacity) upon duration. As in the other figures in this chapter, the numbers in boxes refer to different points within the catchment; these points are illustrated in the catchment map of Figure 5.1. For all the points in Figure 3.7, no significant surface drainage from these points occurred, and the only way out for the water was through the drain. Under these conditions, it is not surprising that duration of flooding depends a great deal upon solids levels and capacity.

Street grading

Common sense and experience show that the shape of the ground in a catchment has a profound influence on the depth, area, and duration of flooding. This importance is hard to summarize in any general way, as every site's topography is different, and the effects of changes in one part of a catchment are not likely to be the same for similar changes elsewhere. Nevertheless, engineers in urban improvement schemes need to think about the effects of changes in street levels upon drainage. Some engineers have explicitly adopted the idea of using 'roads as drains', that is, deliberately designing the roads to act as an integrated major drainage system. The Indian engineer Himanshu Parikh (Parikh, 1990) has been an effective advocate of this approach, and has used it extensively in the slum improvement programme in Indore.

It is not just designers, however, who need to understand the importance of street levels. Site engineers are inevitably confronted with the need to consider changes to design levels because of new circumstances or differing local conditions. Heywood (1994) and Heywood *et al.* (1997) used modelling to examine the effects of changes made by the site engineer on design road elevations in an urban upgrading scheme in Indore. The studies found that these changes resulted in approximately one-third of the runoff being diverted away from the piped system, to flow instead through relatively narrow streets, creating far more of a drainage problem. The changes from design had been provoked by complaints from local residents about the proposed design levels, as their homes' plinth levels had been set based on existing topography.

The site engineer is, as always, in a difficult position when trying to implement the designer's ideal; the lesson is not that such changes must never been made, but only that they can profoundly affect the system's performance.

Both the site engineer and designer need to consider this impact jointly before approving changes from the original levels.

Inlets

In most open drain systems, runoff simply enters along the top of the channel and specific inlets are rarely provided. Such inlet systems are not easily blocked; where one part is blocked, runoff can enter just a bit further downstream. In some systems, a small wall or dike extends above the top of the drain along its length, with only a limited number of points at which runoff can actually enter the drain. Such an arrangement may reduce the amount of debris which finds its way into the channel, but care needs to be taken that the entry points are well-maintained and functional during storms.

For closed conduit drainage systems, inlets play two critical and conflicting roles. They need to:

- increase the entry of runoff, and
- decrease the entry of solids.

If inlets are blocked, poorly located, or set at the wrong level, flooding can occur despite unused capacity of the drainage system. Inlets can easily become the weakest link in the chain of closed-conduit drainage.

Common designs of inlet (Figure 3.8), consist of:

- kerb inlets, where runoff enters a vertical opening in the street kerb
- gutter inlets, where runoff enters an opening in the gutter section, and
- combination inlets, which combine both kerb and gutter inlets.

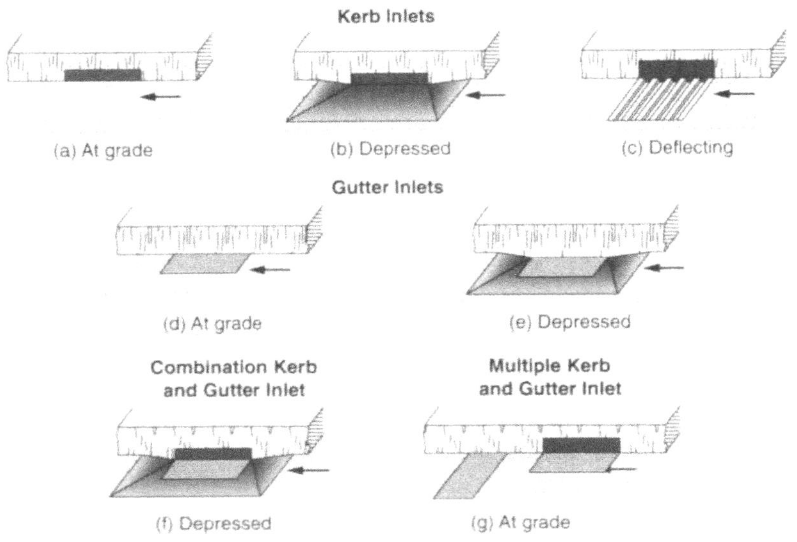

Figure 3.8 *Types of inlets (after Fair* et al., *1966)*

In general, kerb inlets are less prone to clogging than gutter inlets, and are therefore more appropriate in low-income areas where solid waste, construction debris, and other solids are cleared at irregular intervals.

Engineers can ease the entry of runoff by increasing the inlet area, depressing the inlet, and using widely spaced grates in gutter inlets; unfortunately, all of these options also increase the entry of solids. Using narrow grates and inlets, shortening the length of the inlet, and keeping the inlet at (or above!) road level all help to exclude solids, but also hinder the entry of runoff. The compromise is thus not a straightforward one, but the following guidelines seem appropriate:

- *The locations of inlets, like any other part of the system, need to be designed.* In many catchments we studied in Indore, any storm drain manhole, built for whatever reason, was used as an inlet, with the rationale that more frequent inlets meant more effective runoff capture. Unfortunately, more frequent inlets also mean more effective solids capture, leading to long-term deterioration of the network, often with no corresponding improvement in drainage performance. Argue (1986) presents technical guidelines for inlet location specific to South Australian practice; these reflect explicit estimates of the width of gutter flow and the quantity of runoff to be captured.
- *Use kerb inlets rather than gutter inlets where solid waste management is poor*, or inlets are otherwise prone to blockage. Unfortunately, the performance of kerb inlets drops off dramatically with increasing slope along the street and gutter; this may complicate entrance conditions for combinations of storm pipes with 'road as drain' systems. Depressing the inlet and using deflectors can reduce this difficulty.
- *Don't use grates to keep debris out of a kerb inlet*; gully pots (catch pits) are a more effective way of controlling solids. These are a high priority for maintenance.

Some details on inlet performance are presented in Annexe 3-B.

Catchment surface and storage

The effects of urbanization on runoff have been extensively studied (SCS, 1975; Chow *et al.*, 1987; Shaw, 1994). The SCS manual on Urban Hydrology for Small Watersheds (1975) describes one technique among many that have been developed to help engineers estimate these effects on the runoff patterns of an area. While the precise effects are difficult to estimate, the principal effects are in increasing the total volume of runoff, as the surface changes from a less permeable to a more permeable area, and in accelerating the flow of runoff through directing flow in channels and reducing storage.

Effect on runoff volume

Table 2.1 (page 8) shows the variation in runoff coefficients with different types of soils and slopes; the runoff factor can roughly be taken as 1.0 for

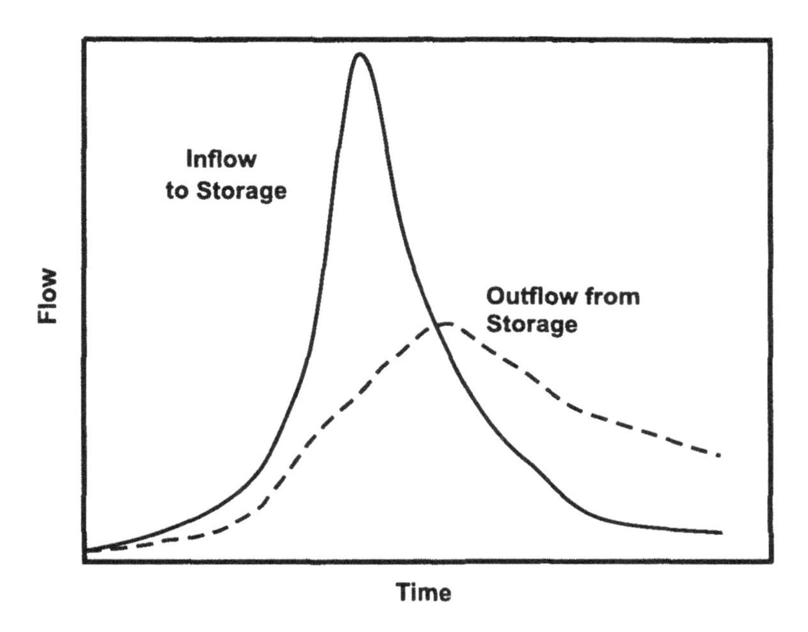

Figure 3.9 *Storage reduces the peak, but not the volume, of runoff*

paving or the areas covered by houses. Urbanization thus has a major impact on runoff through the conversion of relatively permeable areas to impermeable paving and roofing. When assessing the adequacy of an existing system's capacity, engineers thus need to consider not only the surface characteristics of the present catchment, but also how those characteristics will change in the future as open land becomes developed into housing and paved roads. A shift from an average runoff coefficient of 0.4 to a higher coefficient of 0.6, for example, represents roughly a 50 per cent increase in flow. Figure 3.2 indicates that such an increase is equivalent to a fivefold increase in flooding frequency; if the system flooded only once every five years before, it will now flood once a year. Use of Table 2.1 and estimates of both current and future land use and soil cover can offer a rough estimate of changes in flow, and the previous discussion of the effects of capacity offers a guide as to the performance implications of these changes. An example of how to estimate these effects is given in Chapter 7.

Effect on timing of flow

When areas develop, ponds and depressions are filled, and the routes by which runoff reaches the outlet are shortened. This is reflected in a variation in the *time of concentration*, generally defined as the time it takes for water from the most remote part of the catchment to reach the outlet. Paving and drainage works reduce the catchment time of concentration, so higher inflows from shorter, more intense periods of rainfall converge to create a higher total flow.

By contrast, storage in ponds, gardens, green spaces, and flat roofs can substantially reduce peak flows. Engineers often design detention basins for drainage networks as a way of reducing the peak flow downstream of the basin; their effect upon the flow hydrograph is sketched in Figure 3.9. While storage will not substantially alter the total volume of runoff, it counteracts the effects of channelization described above, and allows the same volume to drain over a longer time. Land use changes which fill in large open areas that normally pond during rains will have exactly the opposite effect. They reduce the storage within the system, and thus concentrate runoff discharge into a shorter interval at a higher flow. This can easily increase flow and make flooding worse.

Annexe 3-A: Derivation of frequency and capacity relationships

DEFINE:

T = Return Period [years/event].
f = the frequency of exceedance (rainfall intensities or flows), [events/year],
Q = the peak runoff discharge or drain capacity [flow],
k_1, k_2, and k_3 are appropriate constants, and
C, i, and A are as defined for the rational method, described in Chapter 2 (page 6).

then,

$$T = \frac{1}{f} \quad \text{(by definition)}$$

and, from Kothyari and Garde's relationship between intensity and return period we obtain:

$$i = k_1 T^{0.20} = k_1 f^{-0.20}$$

Substituting this into the rational formula, we obtain

$$Q = k_2 CiA$$

$$Q = k_2 C(k_1 T^{0.20})A$$

which, for a given catchment, can be re-expressed as

$$Q = k_3 T^{0.20}$$

Alternatively, comparing the ratios of flows of differing return periods (or frequencies,) we find:

$$\frac{Q_1}{Q_2} = \left(\frac{T_1}{T_2}\right)^{0.20}$$

$$\frac{Q_1}{Q_2} = \left(\frac{f_2}{f_1}\right)^{0.20}$$

Finally, we can take the last relationship, raise both sides to the fifth power, and compare how frequently two drains of differing capacities (Q_{cap1} and Q_{cap2}) are overloaded:

$$\frac{f_1}{f_2} = \left(\frac{Q_{cap2}}{Q_{cap1}}\right)^{5}$$

Using this relation, if $Q_{cap2} = 0.87 \times Q_{cap1}$, the relative frequency of overload will be $(0.87)^5$ or 0.50. In other words, if a build-up of solids over the years decreased hydraulic capacity by 13 per cent, the pipe will be overloaded twice as often.

Annexe 3-B: Performance aspects of inlets

AT RELATIVELY LOW flows, kerb inlets function hydraulically as weirs. The US Department of Transportation, cited in Argue (1986) report a capacity–discharge relationship for a side-entry inlet as

$$q = 1.66\ l_i\ d_i^{\ 1.5}$$

where q = discharge [in m^3/sec], l_i is the unobstructed length of the inlet section [in metres] and d is the depth of flow above inlet lip [in metres].

Argue reports the following observations on full-scale testing of inlets in Australia:

- inlet length is the most important determinant of capture performance
- 'depression' of inlet lip below gutter level (typical 50–60 mm) leads to significantly improved capture compared with an undepressed inlet.

Argue (1986) and others (Fair *et al.*, 1966, Clark *et al.*, 1977) also recommend that inlets should not be designed to capture *all* gutter flow, but should bypass between 10 and 20 per cent. This is desirable because the hydraulic performance of the inlet is significantly improved by flow entry from all sides (rather than from the upstream side only). This increased performance may reduce water levels in the street for a given flow and the underground investment required as stormwater is routed along the gutter. Figure 3-B.1 (Argue, 1986) gives an idea of the effect upon kerb inlet performance of inlet length and gutter slope; it shows the *total gutter flow* (including that which is bypassed downstream) that can be managed by two sizes of inlet with different lengths and assumed rates of gutter flow capture. Note that doubling the length of the kerb inlet increases the total gutter flow managed by a factor of more than three.

Heywood (1994) presented modelling results for an urban upgrading scheme in Indore which show some general characteristics of inlet performance (Figures 3-B.2, 3-B.3). As in previous figures, the 'nodes' are simply different points in the catchment. Figure 3-B.2 shows that the *depth of flooding* is independent of inlet length at many points in the network upstream of an 'overloaded' (hydraulically surcharged) pipe. At these points, increasing inlet capacity has little effect, as flooding is largely controlled by the surcharge of the pipe and the limited hydraulic capacity of the network. The one node that reflects decreased flooding with increased inlet capacity is near the bottom of the drainage system. Here the pipe network has excess capacity, and

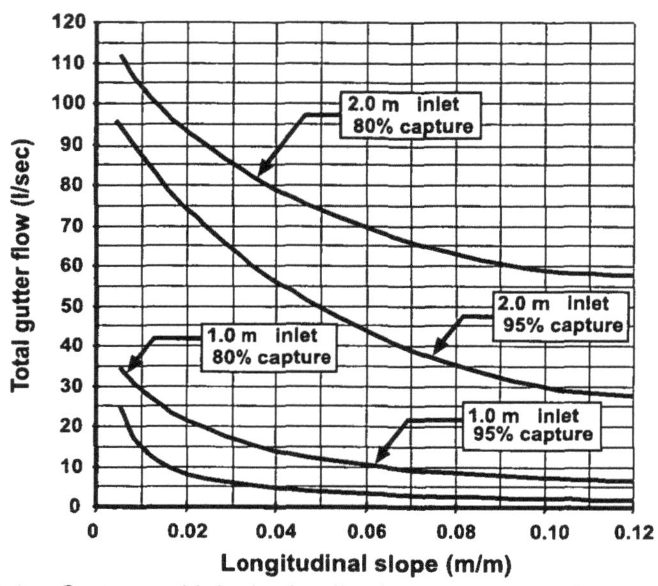

Figure 3-B.1 *Gutter and inlet hydraulic characteristics (after Argue, 1986)*

the inlet acts as a restriction upon the entry of water into the pipe. In addition, increased weir-length throughout the system has increased the fraction of flow going to the pipe, and away from the street. Note that the flooding depths are measured above the *kerb*, not above the road surface.

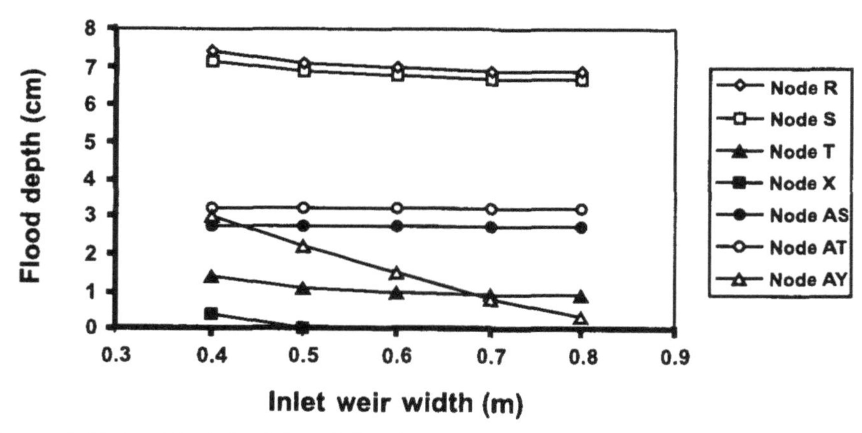

Figure 3-B.2 *Modelled flood depth vs. drain inlet lengths in a design storm (Heywood, 1994)*

Figure 3-B.3 shows that peak inflows to specific inlets are predicted to vary roughly linearly with length, as expected from some design equations. Note, however, that design equations are estimating *potential capacities*; Figure 3.12 shows the prediction that this capacity (at least in this catchment) will actually be *used*, and that increased inlet capacity would thus reduce the quantity of surface flow.

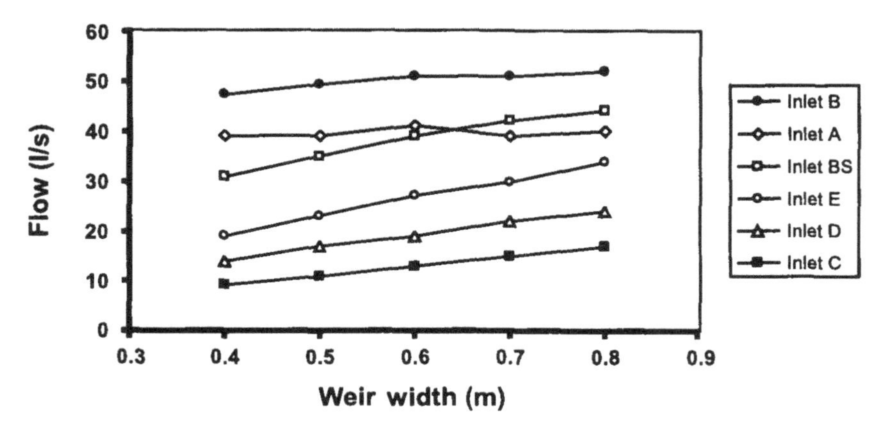

Figure 3-B.3 *Peak inlet flows predicted as a function of inlet lengths (Heywood, 1994)*

4 Drainage evaluation: general approaches

EVALUATIONS ANSWER QUESTIONS. The types of questions to be answered will depend on the purpose of the evaluation. Is management trying to get a sense of the condition of a city-wide system, or are engineers trouble-shooting the problems of a particular catchment? Are major new investments in drainage capacity under consideration, or is the question one of trying to make the most of an existing system through improved maintenance? An evaluation may be wanted to answer several of these questions at once, or it may start off with one question ('Why does this catchment flood so often?') and end with another ('Why is the channel downstream of the catchment always choked?'). This chapter does not therefore offer a detailed 'standard procedure' for evaluation, as the direction is likely to shift and change as more is learnt about the system. Instead, the chapter discusses some broad approaches (depending upon the purpose of the evaluation), and then considers evaluation methods in terms of the questions they can help to answer. These are summarized in Table 4.1 along with relevant page numbers. Chapter 2 pointed out that an evaluation could attempt three types of measurements, depending upon the questions being asked, and the available resources:

- performance measurements
- performance indicators
- process indicators.

While this manual will describe measurements of all three kinds, it is focused largely on performance indicators and process indicators; these tend to be both simpler to perform and more useful for field practice than rigorous performance measurement. Performance measurement is, however, important for research to link performance indicators to performance, and is therefore briefly described for those with such an interest.

System-wide evaluation

An evaluation of the entire drainage system of a city may be needed for a variety of reasons. It is often required for funding or planning decisions about where and how much to invest in new facilities, e.g. the preparation of a traditional drainage master plan. Such an evaluation needs to take place on

http://dx.doi.org/10.3362/9781780446059.003

Table 4.1 Drainage evaluation questions and methods

	Page Nos.
What is the catchment like?	48
Topographic survey	48
Development of working catchment map	49
Cover survey	52
Is flooding a problem in this catchment?	54
Resident surveys and interviews	54
Resident observation gauges	59
Chalk gauges	59
What kinds of flows can be expected?	65
Catchment cover	65
Rainfall data analysis	71
What is the approximate capacity of the conduit system? (design, as-built and actual)	81
Survey of conduit system	86, 94
Dimensions, levels, slopes	94
Checks for blockages	100
Solids deposits	106
How does the drainage system behave in practice?	118
(Check catchment boundaries, inlet blockage, bottlenecks, flooding, standing water)	
Wet weather surveys	119
Are there construction problems in the drainage network?	94
Open conduit	
Inspection	100
Closed conduit	
Standing water checks	101
Lamp-mirror survey	101
Manhole inspections	104
Are inlets restricting drainage?	95
Inlet survey	105
Maintenance survey	112
Is maintenance a problem? Can it be realistically improved?	106
Drain solids surveys	106
Solids build-up surveys	107
Inlet maintenance surveys	112
Drain cleaning observation	113
Drain cleanings piles surveys	115
Solid waste monitoring	116
What is blocking the drain?	106
Visual observation	
Solids sampling	110
Particle size analysis	110

several levels. In practice, such studies often focus on the 'primary' infrastructure (e.g. main drains and pump stations) and pay less attention to the secondary or tertiary network (minor collector and street drains) that feeds into the primary system. The focus on primary infrastructure is natural because of the complexity of drainage, the need to keep the scope of the study within bounds, and the need to gather data about the common components that affect everyone. The drawback of this approach is that many drainage problems occur upstream of the primary system. In such cases flooding may not be reduced through improvements to the primary system, because the secondary or tertiary system is inadequate. It is therefore necessary to evaluate on a variety of levels.

Who will perform the evaluation? If you are in municipal government, then this question is not always an easy one. On the one hand, nobody knows the system as well as those who work with it every day. External consultants often appear expensive relative to in-house salaries, and they will spend quite a bit of time (and money!) becoming familiar with information that your staff already understand. On the other hand, the true cost of doing the job 'in-house' must be carefully considered; these include delays in other work from committing time to the evaluation, and an evaluation of lower quality because of the need to stay on top of other demands. No matter who is responsible for conducting the evaluation, those who work with the system must be *involved*, or far too much knowledge will be lost, or rediscovered the hard way. Thinking clearly about whether or not to use consultants is essential in setting a realistic scope of work for the study; the answer depends on the type of questions you are trying to answer.

A system-wide drainage evaluation should probably involve the following stages, although they may not be carried out in strict sequence:

Gather background data

Talk to those experienced with drainage problems in the city Many of these will be within the drainage authority, but some may have moved on to other jobs or retired. What do they see as general problems, what do they see as specific problems, and what do they see as potential solutions or improvements? Mark on a map those areas that they perceive as particularly troublesome. Try to determine the basis of their opinions: is it direct observation, or is it a feeling acquired over years of experience, observation, and listening to employees and the public? In addition, talk to residents' associations and NGOs and look through newspaper archives to get a sense of problems. In many cases, these will confirm what the drainage authorities have already indicated. In some cases, they may indicate different problem areas. For one thing, drainage authorities are often responsible only for the areas they officially serve; by definition, they are often not responsible for 'informal' or 'illegal' areas outside their jurisdiction. Furthermore, municipal drainage authorities are often influenced by complaints (as they should be!). Some communities are better able to organize pressure on municipal authorities than others, and this may distort the perceptions of drainage problems.

Gather and review existing topographical and drainage system data and reports These are often scattered among a range of offices: for example, between the Planning Department, the Public Works Department, the Development Authority, and the operations and maintenance staff. Such a review of existing data will almost certainly reveal important gaps. In many cases basic topographical data will be inadequate for even broad analysis; in others, basic system data (e.g. maps, sections, and slopes of drains) will not be available. Make copies of whatever data are available and assemble them in a single place. Even if no additional data are collected, such a comprehensive gathering of scattered data is a valuable resource for any further work in drainage, by you or by somebody else.

Identify preliminary questions What are the specific questions that seem the most important to resolve? What are the hypotheses (tentative explanations for problems) that are to be tested, and how will they be tested? Is attention restricted only to principal drainage routes, or is the study concerned about flooding in specific areas? Decide if the study must look at secondary or tertiary drainage, as well as the main system. If so, it may be necessary to identify 'typical subcatchments' for study which cover a range of slopes, extent of paving, location (e.g. near or far from an outlet) etc. Alternatively, the study may focus only on specific 'problem catchments' (see the next section, page 45), and identify their problems.

This is one of the most important stages of the work, and yet one that is easily neglected. It is relatively easy to collect a lot of data, but much more useful and efficient if these data are selected to answer specific questions. You probably can't answer all the questions you would like within your time and budget, so you need to work out the questions which are most important, and most likely to be answered through evaluation.

Identify information gaps What additional data need to be gathered? Who could have them? Do they have them, or do the data have to be gathered from scratch? The study will probably need more information for a complete analysis than can possibly be obtained in the time and budget available. The object is then to achieve a useful evaluation within time and budget constraints, rather than a complete and comprehensive evaluation that is only half-finished because only half the data can be collected. Information requirements therefore need to be prioritized. It may be better to get some data on a wide variety of topics; these can suggest some answers to many questions that may be checked later. Alternatively, it may be better to do detailed analysis of a single small area. The answer depends upon the problem at hand, but before much work is done, it is essential to review the information needs against the time and resources required to fulfil them.

Perform field work

Field check your most critical background data Do the available drawings of major drains show cross-sections and levels? Most of the available plans and drawings will be 'design' drawings (prepared to show how the designer thought the system *should* be built), not 'as-built' or 'completion' drawings (prepared to show how the system was *actually* built). These can differ radically because of unforeseen problems that arose during construction (e.g. buried pipelines that must be crossed, a shortage of materials, complaints from residents leading to re-routeing of channels, etc.). Beside field checking the difference between design and as-built dimensions, it will be necessary to look into blockages in either open or closed channel systems. Chapters 9 and 10 describe aspects of this work.

Gather additional data The type will depend upon the questions to be answered. Performance data, for example, can be gathered only during storms, whereas performance indicators and process indicators may be gathered during dry weather. If you want to understand the impacts of frequent flooding, you will need to listen carefully to residents' perceptions of what happens when it floods, and of which aspects are most troubling. Residents can also help to gather performance data to answer some of these questions, as they did in Indore. On the other hand, the study may be interested only in performance or process indicators, such as solids levels in drains, or blocked inlets.

Residents may have theories as to the causes of flooding. By the end of the evaluation these theories may be proven right or wrong, but they are certainly worth knowing and understanding for two reasons. First, residents see and experience more flooding in the area than anyone else, and may well see much that is not obvious to an engineer. Secondly, if there are big misconceptions about the causes of flooding, you may want to clarify them at the end of your work.

Analyse the data

Look at what has been gathered and see what stories it tells. If possible, try to look at it while you are collecting it. You can do this by setting aside a regular period in the day to look at what was gathered the previous day to see if it makes sense. For example, in a topographic survey, it is important to check the errors during the survey period, so that if data must be reconfirmed, it can be done right away; going back months later is much more difficult. See if such 'field analysis' suggests other questions you can check easily, but do not get carried away looking at too many new questions!

Analysis can take on many forms. In some cases, for a major investment programme, it may be worthwhile to use computer packages specifically designed for hydraulic analysis. Computer models also have the advantage that they can make data management (and therefore collection) more systematic; they have the disadvantage that once data are entered in the model, people tend to believe their validity without checking them. Annexe 8-A outlines some of the issues in considering the use of software for drainage analysis.

Many other forms of analysis can be done without computers, or with very simple spreadsheet packages. These include checks on the build-up of solids, or the percentage of blocked inlets, or the average fraction of encroachment upon open drains. These simple numbers also tell stories, and indicate possible weaknesses in the system. Chapters 8 through 11 describe these approaches in greater detail.

Write up the findings

You may have learned a lot about the system, but you need to write it up for others to learn and use the findings! In addition, you will probably gain additional insight as you try to explain it to somebody else. For this reason, it is a good idea to draft an outline of the report's structure at the outset of the study, revise it as you gather data, and then again as you analyse it. These outlines can be shown to various people as you develop them, so that they can comment on items they think are missing, or which are not very relevant. Table 4.2 is a draft table of contents for a system-wide evaluation that might be done by a consulting firm. It may not be appropriate for your area, or you may not have the resources to answer all these questions. It may nevertheless serve as a starting point from which you can define the type of investigation you or others may undertake.

Evaluating a specific catchment

The process is broadly similar to that for system-wide evaluation; the differences are largely in the scope and detail.

Gather background data

Talk to those familiar with the area's drainage These include not only the engineers in charge of the area's drainage, but also residents and maintenance staff. Ask about the problems experienced in the area, and seek out opinions as to the causes.

Gather existing data Check to see if there are earlier studies (including design drawings) of part or all of the area, and gather whatever maps and topographic data are already available. These may be scattered among a variety of sources, e.g. housing authorities, drainage authorities, planning authorities, maintenance departments, etc. Copy these data, and gather into a single place.

If possible, define the catchment. The catchment consists of the entire area that drains into the system under study. Determining this in practice is often more difficult than it seems at first, but the task is worth the time. Often the available maps and drawings are of either specific contract sites or housing developments, rather than of the catchment, which means at best piecing together maps of several different areas. At worst, substantial portions of the catchment are unmapped, so that boundaries are indeterminate without a good field survey.

Table 4.2 Table of contents for system-wide drainage evaluation

Executive summary
- Background and purpose of evaluation
- Main questions
- Main points of the entire evaluation, with particular emphasis on the recommendations

Introduction
- Background and purpose of evaluation: who requested and why?
- Structure of document
- Acknowledgements

Approach to evaluation
- Prior work and its limitations: why evaluation is needed
- Identification of main questions
- Identification of specific (geographic) problem areas
- Identification of general (topic) problem areas, e.g. system condition or maintenance
- Approach of study: methods adopted and why
- Work done as part of this study

The main service area(s)
- Definition (boundaries) of service area and main catchments
- Existing drainage network (type, capacity briefly described)
- Existing topography and development, any foreseen changes
- Patterns of existing and future urban development, soil cover
- Maps, including topography, cover types, development patterns, existing drainage system, main catchments, and identified drainage problems

Hydrology and drainage requirements
- Local hydrology (rainfall patterns)
- Maps showing secondary subcatchments
- Explanation of runoff estimation technique
- Current runoff patterns and estimated flows
- Future runoff patterns given foreseeable future development
- Identified problem areas, described by type and extent of problem

Evaluation of primary drainage system
- As-built capacity (i.e. assuming clean and in good repair) comparison of capacity to requirements
- Quality of existing maintenance
- Structural condition of drains and mechanical condition of pumps
- Actual capacity (allowing for structural, mechanical, and maintenance defects)
- Comparison of actual to as-built, and to requirements
- Bottlenecks
- Recommendations

Secondary drainage systems
- For each secondary drainage system:
- Description of any problems experienced
- As-built capacity (i.e. assuming clean and in good repair)
- Comparison of as-built capacity to requirements
- Structural condition of drains and mechanical condition of pumps
- Quality of existing maintenance
- Actual capacity (allowing for structural, mechanical, and maintenance

defects)
Comparison of actual to as-built, and to requirements
Bottlenecks
Outline of major system; what happens when secondary system floods?
Recommendations

Tertiary drainage systems (sampled)
Basis for sampling, and description of sampled areas
General description of commonly used approaches to tertiary drainage
Structural condition of sampled systems
Quality of maintenance of sampled systems
Major drainage systems: what happens when tertiary systems flood?
Recommendations

Unserved areas
Description of current drainage patterns and consequences
Outline of drainage options, including maintenance implications
Recommendations

Maintenance issues
Solids cleaning
- Extent of blockage
- Types of blockage (particle size analysis)
- Frequency of cleaning
- Process of cleaning and disposal of solids
- Existing programme for cleaning (driven by complaint, or periodic?)

Repair
- Extent and type of structural problems
- Process of repair
- Existing programme for inspection (driven by complaint, or periodic?)

Recommendations

Institutional aspects
Who is responsible for what? (e.g. geographic areas, primary/secondary systems, construction/maintenance)
Overlaps in responsibilities and consequences (primary, secondary, tertiary)
Gaps in responsibilities and consequences (primary, secondary, tertiary)
Recommendations

Conclusions
Review of recommendations
Prioritized list of improvements
Further work

Identify preliminary questions In problem areas, some of the following questions arise very frequently:

- Is flooding due to inadequate capacity downstream, is there a local capacity problem, or both?
- Does all the area drain to the drainage system, or are there pockets that do not drain, where water ponds for hours or days?
- Where does the water go during a flood? How quickly?
- Are there specific bottlenecks, or are there capacity problems throughout?

- How blocked is the existing system? How badly are channels, inlets or pipes blocked? What is blocking these systems? How easy is it to clear them, and will they only become blocked again?
- How developed is the catchment? Will there be more runoff in the near future as the area develops, and will this extra runoff exceed the capacity of the existing system?

Other questions will present themselves from your discussions with engineers, maintenance staff, and residents.

Identify and plan your data collection Once you have framed your initial questions, you can then plan your data collection. The rest of this manual outlines the experience of the Indore Drainage Evaluation Project in applying a number of evaluation methods to answer questions about specific catchments. Some data are essential to gather for any study (e.g. catchment definition), while other data may be needed only to investigate specific issues. You will need to consider the weather in planning your work. On the one hand, observation during storms shows many things that cannot be understood during dry weather (Chapter 11); on the other hand, field work during rain is difficult, and requires good preparation.

Perform field work

Field check your background data Background data supplied by others may be incorrect. Design drawings often differ from what has been built, and in any case the condition of any drainage system for blockage or structural damage needs to be double-checked (Chapter 10). You must walk the catchment boundaries to confirm the tributary area; in flat areas it may be best to do this during a storm to confirm where water flows.

Gather additional data By now you will have a clear idea of the most important questions to answer. Check Table 4.1 to see if methods are available to answer these questions. Not all the data needed will be engineering data; residents will be able to share their observations of flooding and performance issues.

Analyse the data

As with city-wide analysis, there are a number of ways to look at data, and not all of them require a computer. Simple data can be plotted by hand, and much basic hydraulic analysis can be done with a calculator and a pencil and paper. As in city-wide analysis, it is best to analyse the data as they are collected, so that 'odd' data can be checked, and unusual trends can be picked out early for more detailed investigation. When planning the work, allow plenty of time for analysis; collecting data that are never used is a frustrating waste of time and money!

Write up the findings

As mentioned for city-wide analyses, it is worth drafting an outline of the report near the start, to clarify what data are needed to answer all the questions behind the evaluation. The outline can be updated during the evaluation, as more is learnt about the problems of the catchment. It may be easiest to start with the outline of Table 4.2, and modify as appropriate for the problem at hand.

5 Studying the catchment

Topographic survey

TOPOGRAPHICAL DATA ARE needed for the following:

- to define the catchment area, and its subcatchments
- to determine the directions of flow in both natural and artificial systems
- to compute flows and capacities in combination with other data
- to identify bottlenecks or control points, which determine drainage patterns for the whole area.

Any evaluation based on inadequate topographic work will miss some of the most important constraints on the drainage system; there are no substitutes for good topographical data, and no shortcuts to the work needed to obtain them.

Level of accuracy

It is impossible to prescribe universal requirements of accuracy. At the very least, topographic data should be at least as good as those that are normally used for design. In many cases, 'design' consists of simply specifying a lining for an existing ditch which flows by gravity, so that many drains are 'designed' without accurate topographical information. To evaluate the drainage system, however, someone must estimate its capacity, and will therefore need to know slopes and levels. If the catchment is steep, topographic accuracy is less critical; in flat catchments, determination of both the catchment area and the slopes depend more critically upon good topographical data. Standard textbook figures (Malcolm, 1966) for ordinary levelling accuracy in favourable conditions are approximately 0.05 per mille (0.05/1000) or 5 cm/kilometre distance. In practice, the error actually varies as the square root of distance, so that a four kilometre distance should have an error of only 10, not 20 centimetres.

Working with closed, rather than open, traverses is more important than precision. A closed traverse is one which either returns to its starting point, or in which both endpoints are already accurately known from previous surveys. Closed traverses permit you to know what error you actually have, and you can then decide if it is acceptable for your purposes. Where open traverses are used (in which measurements are not tied in to known elevations and locations), there is no check at all on errors in field measurement.

It is good practice to establish a series of benchmarks throughout the catchment with which to close traverses. Benchmarks must naturally be determined with greater precision than other levels, as any errors in them will be compounded in any measurements derived from them. Essential details of surveying are covered in a number of basic texts (Malcolm, 1966; Breed and Hosmer, 1977; Stern *et al.*, 1983).

Data to collect

On the general catchment surface it is worth obtaining the following data:

- spot elevations and locations at road junctions
- elevations of road and roadside (is adjacent land above or below the road?)
- elevations and general topography of open land
- low and high spots.

In areas where flooding is experienced or expected, it is important to gather data on housing plinth levels and the levels at which water can enter the house. Topography of the catchment surface is at least as important as that of the artificial drainage system, as water will flow over the surface whenever a flood occurs. In addition to gathering specific data at points of interest, the topographical data should be collected with a view to sketching ground contours at suitable intervals, to show the general lay of the land, and to estimate the direction of flow. In flat areas, it may be difficult to produce meaningful contours, and *it is better to leave out uncertain contours than to mislead by sketching them in.*

Later analysis will also require levels and dimensions for key drainage network points:

- invert (inside bottom) elevations of drainage conduits at principal inlets, outlets, and junctions
- dimensioned conduit sections
- inlet levels
- inlet lengths.

Note that much of the drainage network data can be collected separately from the topographic levelling survey, provided measurements can be linked. For example, if the levelling survey takes 'an elevation at a manhole', the point at which this elevation is taken should be marked on the manhole frame. A separate team with only a tape can then measure the distance down from the frame to the invert of the channel. Communication between the two teams is, however, essential, or else the taping team may measure from a corner of the frame that might be substantially higher or lower than that used by the levelling team. This is described in further detail in Chapter 9.

Analysis

Catchment maps (Figure 5.1) and profiles are two particularly useful outputs.

Catchment maps It is often helpful to have both a large-scale map for detailed office work, and a smaller size one (preferably A4, and no bigger than A3) for future field work.

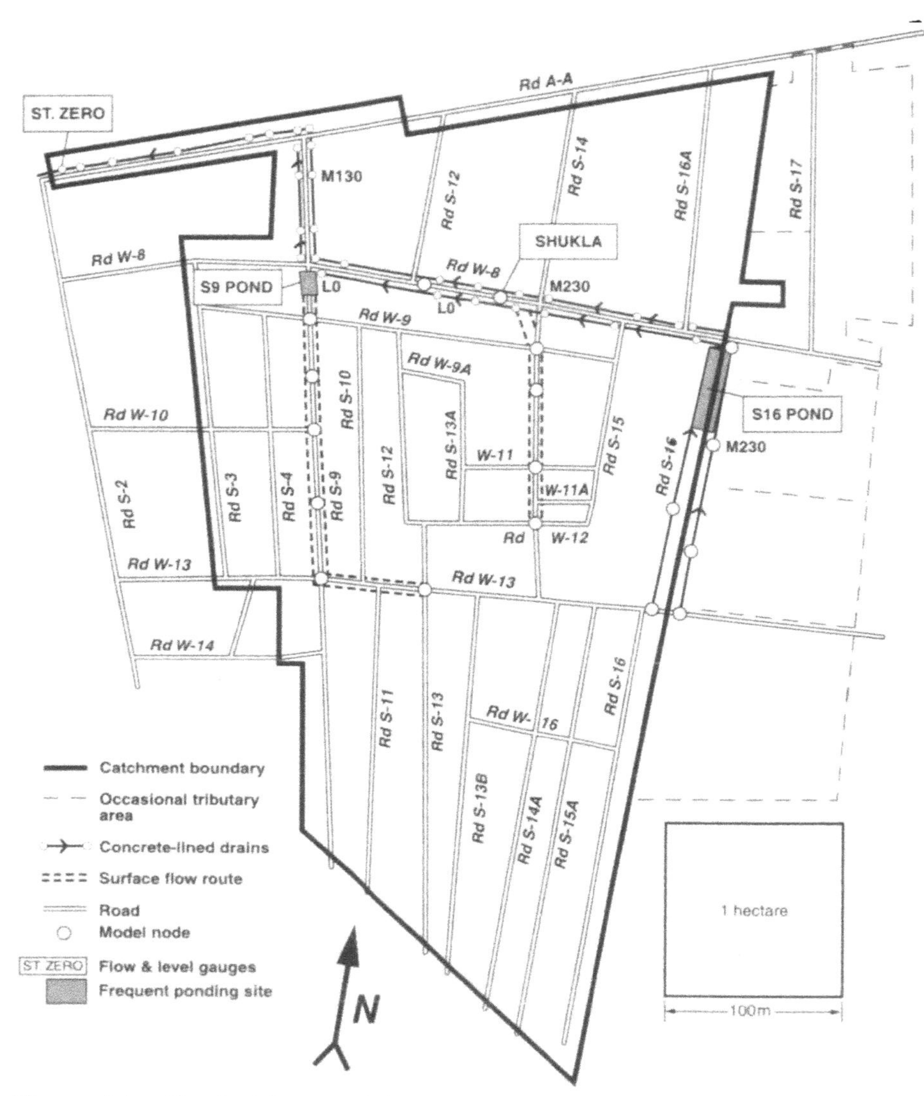

Figure 5.1 *Pardeshipura catchment map*

This working map should show:

- the catchment boundary
- drainage network location and relevant link and node (conduit and junction) numbering
- street identification
- a scale
- a North arrow.

The working map should, at the very least, allow workers to know where they are when performing field work. When photocopied, the working map can thus serve as a baseline for additional field work (e.g. for surface cover surveys, rubbish surveys, or during wet weather observations). Street identification is very important, as in many low-income communities, streets have no formal names or means of identification. If such identification is unavailable, it is helpful (if residents don't object) to paint identifying street names or numbers at intersections during the survey. The working map will also be useful for writing up the evaluation, and explaining findings to others.

Profiles Once channel or pipe sections and levels are gathered, profiles can be plotted up that show the drainage network in section. These are invaluable as a graphic description of the slope of the system, and often pinpoint problem areas, where subsidence or poor construction practice has created dips in the system, where silt and debris accumulate. It is worth doing this for the road system as well, to identify how it will direct the flow of stormwater, and how it will behave as a 'major drainage system'. Figure 5.2 shows a profile of a surface water drain, developed from the type of structural survey described in Chapter 9; note the 'valleys' in the pipe where sediment is likely to be trapped. Such a profile cannot be based simply on 'distances from ground level', but must use accurate survey work to allow for variations in ground level.

Defining a catchment

A catchment is simply the area that contributes runoff flowing past a specific point of interest. Thus the overall catchment is defined as the total area

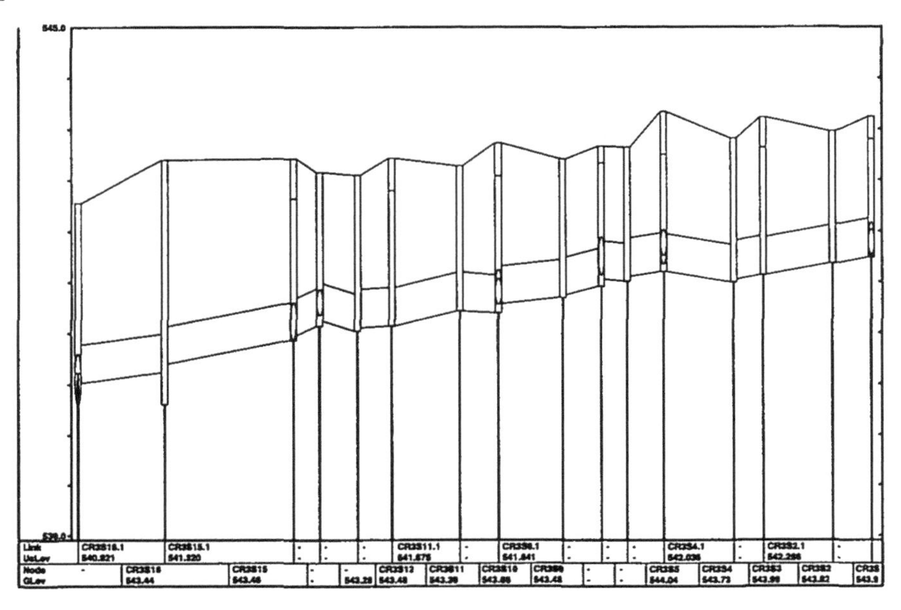

Figure 5.2 *A profile of a 450 mm drain in Indore*

contributing to flow at the outlet of the entire system; subcatchments are defined for specific points further upstream along the main or tributary drains. Defining a catchment seems a straightforward task, but there are often both practical and conceptual difficulties. One practical difficulty is that catchments are defined by topography and drainage networks, not by convenient ward, district, or contract site boundaries; this means that it will probably be necessary to patch together several maps of different areas (often drawn to different scales!) to cover the catchment, and there may well be some portions that are unmapped. The conceptual difficulties arise in flat areas, where boundaries are unclear, and often determined as much by small drainage paths as by topography. In Indore, for example, we noted that the catchment area could actually vary with storms; in small storms, several catchments could drain independently to a collector drain, but in larger storms, overflow from one would spill into an adjacent catchment, thus enlarging the catchment area in later stages of the storm.

These difficulties are real, but can be managed well enough for the purposes of practical evaluation. The main principle is to use the understanding of the topography and drainage network to define the areas contributing flows to different portions of the network. Once the outlines are sketched on a map, it is important to walk the boundaries during a storm, to see if runoff is indeed flowing in the expected way; sometimes alternative drainage routes (e.g. small drainage conduits, back alleys, or routes over properties) emerge which may divert runoff away from the natural slope of the land. In particular, note the levels of roads in the catchment relative to the adjacent terrain. Are they at the same level and therefore simply part of the catchment? Are they set above the surrounding land, so that they act as subcatchment boundaries? Or are they set below the surrounding land, so that they act as drains for the adjacent terrain? This will vary from site to site within the catchment, but noting the difference between the road levels and surrounding ground levels is critical to understanding how the catchment behaves.

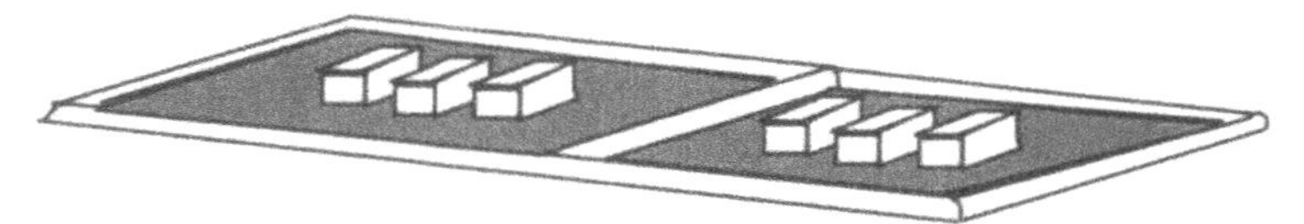

(a) *Roads above adjacent land act as catchment dividers*

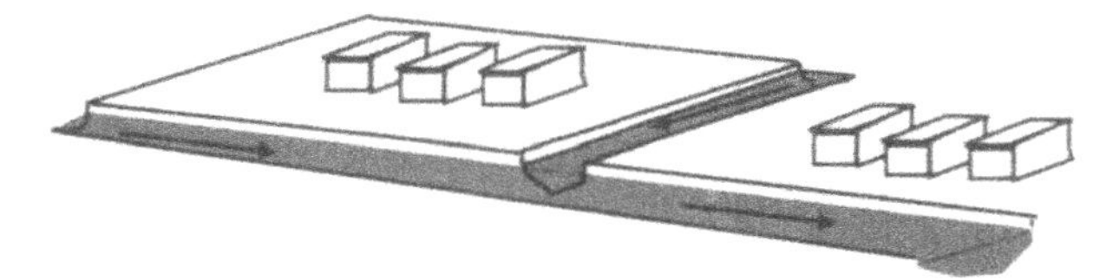

(b) *Roads below adjacent land act as drainage routes*

Figure 5.3 *Effects of relative height of roads and adjacent land*

Surface cover survey

In addition to data on levels, evaluation will require study of the surface cover in the catchment area in terms of runoff coefficient (see Chapter 2). To keep this work manageable, it is a good idea to reduce the number of cover types that you consider. In Indore, we kept the number down to about four types. There is no point in distinguishing between catchment types that have very similar runoff coefficients. In an urban catchment it will be necessary to consider the nature of at least three types of areas:

- blocks or areas surrounded by roads with houses and buildings
- roads
- open spaces (waste lots, or empty housing lots).

It may be easier to tackle a survey of each of these separately. Good maps will make this work much easier, so the cover survey should follow the topographic work. With good maps, the fraction of the catchment taken up by roads can be estimated as a desk study, with spot checks in the field to ensure the quality of the mapping. The main field work will then be to determine how much of the open or developed land is paved or covered with housing, and what other kind of cover is present.

In the housing areas, it will be necessary to walk around or through the block to estimate the fraction that is paved, and the nature of the permeability of the other areas. As an extreme example, consider a catchment in a flat humid area, and the runoff coefficients shown in Table 2.1. If buildings take up about 30 per cent of the surface area of a block, and the rest is sandy soil, then the C-value is 0.30×1.00 (for the buildings) $+ 0.70 \times 0.05 = 0.34$. If the soil in this example were clay, however, the computation would be $0.30 \times 1.00 + 0.70 \times 0.55 = 0.69$. This shows the importance of the cover survey, as these two areas differ in their projected peak flows by a factor of two!

It is worth developing distinct teams for this work, who can cross-check each other's work for consistency. There is a subjective element in estimating areas and soil types, so it is worth checking in the early stages of the survey to see if individuals are able to achieve fairly consistent results on the same sites. By the end of the survey, you will be able to work out the area of a number of different types of cover for the total catchment and the subcatchments. This information is needed to estimate flows, as described in Chapter 7.

6 Assessing flooding as a problem

THIS MANUAL HAS defined drainage performance as the depth, area, duration, and frequency of flooding. In this chapter, we discuss five methods for assessment of drainage performance and the nature of flooding as a problem:

- resident surveys of the community about flooding
- direct observation during storms
- resident gauges which are local landmarks (e.g. telephone poles) marked as scales by which local residents can observe flooding depth and duration
- chalk gauges which are chalk-coated metre sticks, installed in protective boxes into which flood waters may enter
- electronic level gauges (such as by a recording pressure transducer) which record the water level in a drain.

The main advantages and disadvantages for each method are shown below in Table 6.1. In general, resident surveys and direct observation are more than adequate to describe the problem, and together represent reasonable performance indicators. The other methods are more direct performance measurements, but are also more difficult to manage, and are unlikely to be of much practical use except as a research tool. This chapter will therefore indicate only a few basic points from our experience in trying to map flood-prone areas in Indore.

Resident surveys

People who have lived in the community for several years know a lot about the risks and extent of flooding in their area. They will definitely remember occasions when water flooded into their homes, and they will have heard from friends and neighbours about how much worse (or better) the problem is a few blocks away. You can get an idea of which areas are worst affected by simply talking to people and asking around. A number of books and articles on both community planning and participatory evaluation (e.g. Feuerstein, 1986; Goethert and Hamdi, 1988) give hints and examples about how best to work with communities in assessing a wide variety of problems, which could easily include flooding. In addition, if you want to do more detailed work on community perceptions of flooding, you may wish to consider applying a variety of 'qualitative' research methods, which don't yield numerical results (e.g. average depth of flooding), but rather insight into how people view

http://dx.doi.org/10.3362/9781780446059.004

Table 6.1 Flood assessment techniques

Method	*Advantages*	*Disadvantages*
Resident surveys	Low cost Fairly quick to organize Can be done in dry weather Can cover a large area	Depends on memory Most reliable on depth; less reliable on frequency and duration More subjective than other methods
Direct observation during floods	Can observe all of problem in context Essential for other parts of evaluation (Chapter 11) Useful check on, and complement to, other data	May be seasonally limited Difficult to be at right place at right time: How do you catch the peak? Unlikely to observe full hydrograph Can only cover limited area Dependent upon weather
Chalk gauges	Low cost (relative to electronic sensors) Captures floods at any time Useful check on resident gauge Can cover large areas	Only records maximum depth Only valid during season of observation Requires frequent visits Siting is critical
Resident gauges	Low cost to researcher Can involve community in study and serve as basis for discussion of flooding	Requires time and effort from residents, who are volunteering services Can't record floods if away or asleep Hard to read at night
Electronic sensors	Accurate and reliable 24-hr/day coverage Can produce good hydrographs	Expensive Can only cover limited number of sites Require inspection, maintenance, battery checks, and external verification Requires protection from vandalism

flooding. *This is my Beautiful Home* (Stephens *et al.*, 1994) reports on one of our Indore studies that employed qualitative research methods and yielded some important results about how residents viewed problems of flooding and drainage (as discussed in Chapter 2, page 16).

Avoid 'leading' questions

The answers people give to questions depend a lot on the way the questions are asked, and how the questioner presents him- or herself. It is important that each question be as open and neutral as possible, allowing the person being interviewed to express themselves as freely as possible. If it is clear that the questioner is looking for a certain type of answer, two types of problems can arise: (1) people try to be 'helpful' and give the answer the interviewer appears to want rather than the truth, or (2) people may resent the assumption that the interviewer knows what the answer is, and may then try to mislead or surprise the interviewer. Either way, the answer is not helping you to understand the truth. Consider the following two introductions to a discussion about rainfall and flooding with a community resident:

A. 'We're engineers from the drainage department, and we're trying to decide where to use some of our money to improve drainage in the city. We understand flooding is a severe problem in this area, and many people have complained to us about how bad it is. Tell me, do you think flooding is a problem in this area?'

Every statement in this introduction may be true, but it can easily prompt exaggerated answers from a resident. If the questioner is viewed as the representative of an authority or organization with money to spend on drainage, and if it is implied that neighbours think they have a drainage problem, the respondent is more likely to agree, if only to ensure peace with his or her neighbours.

B. 'We're students from the university, and we're trying to learn about rainfall and its effects on different parts of the city. What happens here when it rains?'

Here, the interviewer represents a relatively neutral body, and is simply expressing curiosity, without a particular 'position' towards flooding as a problem. The question is an open one, which is not simply answered by Yes or No, and to which there is no obviously 'expected' answer.

Ask more than one person

If just one or two people are asked, they will know some parts better than others, and their response may not be representative. Sometimes only community leaders are asked, on the grounds that they know about 'the whole community'. Here there is an additional risk that the answer may exaggerate a problem, if it is suspected that action to improve drainage problems may result from the answer. It is also important to establish where respondents spend most of their time; if men work outside the area, and women spend more time around the home, then women may have a clearer understanding of the problems of minor flooding. Checking with many people tends to give a range of answers with some very high or very low replies, but most grouped around a consensus. In addition, some discussions in groups may be helpful to identify which opinions are unusual, and which ones appear to be shared by the group.

Try to be specific

It is probably better to ask first about the previous year's flooding, rather than asking 'in general, how high does water rise?'. Asking 'in general' is really asking people to do a very complicated 'averaging' exercise in their heads! By contrast, people often remember last year's level quite clearly, and the highest level they have ever seen. Similarly, interviewers need to be clear about the reference level of the 'depth' of flooding; does it refer to the middle of the street, the household plinth, etc. It is best if the residents show the interviewer a specific place where they remember the high water mark, rather than discussing general depths such as 'water was hip deep'.

Special care needs to be taken if some or all of the team is working in languages different from those of residents. 'Flooding' may occur in one language when small puddles form in the road, while in another, it may imply water overflowing the kerb.

Retrospective flood surveys in Indore

In two Indore catchments, we asked a number of residents and local merchants about the depth and duration of flooding experienced the year before, and marked on maps the maximum depths reported at their respective locations. Such surveys will be far from precise, although in some places people will have a very precise marker as to how high the water rose.

While there were one or two unusually high or low responses, most were fairly consistent, and the surveys confirmed the portions of the area which had the most significant problems. We concluded that such surveys do help to establish a general idea of the extent of flooding in the community, and are quickly and easily done. Residents may volunteer the highest level, or they may remember a particular level (e.g. the kerb) being overtopped an approximate number of times. It is best to let residents choose the way to express height, as they may remember, for example, when water came over the plinth, but not over the kerb. Table 6.2 shows one sample survey sheet.

Our results were also qualitatively consistent *between* catchments. One catchment that had experienced virtually no significant flooding, effectively reported this, while another, which had been subject to extensive flooding, yielded results which were consistent with our own informal observations of the catchment the year before.

Direct observation

Walking around in a flood is one of the best ways to see what happens when it floods: which areas are affected, how badly, and the natural drainage routes. For this reason, Chapter 11 describes the objectives and approaches to wet weather observation. However, as a single method of assessing the severity of flooding this approach is demanding, frustrating and limited because:

- you may not be able to start until the rainy season
- you cannot be everywhere all the time
- you can easily miss the most important storm, or the most important part of it
- you may have many other evaluation objectives to satisfy during a storm (e.g. catchment definition, observation of flow patterns, etc.; see Chapter 11).

Table 6.2 A form for recording surveys of residents about floods

RETROSPECTIVE FLOOD SURVEY

Site: **Date:**

Surveyor:

(1) *Street ID*	(2) *Section*		(3) *Location*	(4) *Height of flooding*	(5) *Above what base?*	(6) *Base height above road*	(7) *How often to this height?*	(8) *How long at this height?*	(9) *How long to drain?*	(10) *Remarks*
	From	*To*		*(cm)*		*(cm)*	*(times last year)*	*(mins, hrs, or days)*	*(mins, hrs, or days)*	
S2	A	W8	NE corner of S2 & A	15	Road	0	15	10 min	24 hrs	Near station zero; water overflows from drain
			Junct. S2 & W8	35	Kerb	+15	6	20 min	6 hrs	
	W8	W10	20 metres West of S2 & W10	40	Road	0	10	1 hr	2 hrs	water overflows from drain, collects in road
			Junct S2 & W10	45	Road	0	12	30 min	3 hrs	

Column explanatory notes.

(1) Each street is assigned an identification number or code during catchment mapping.
(2) 'Section' shows the portion of the street where data are being collected by identifying the cross-streets 'From' and 'To' which define the section endpoints.
(3) The location on the street section to which data apply.
(4) Highest level of flooding in previous year
(5) Flooding is often referred to differing 'base' levels, e.g. kerb, plinth level, etc.
(6) Difference between road level and 'base' level at this point.
(7) How often did it flood to this level last year?
(8) Approximately how long did it stay at this height?
(9) How long did it take for the street to drain at this point?
(10) Any other points worth noting.

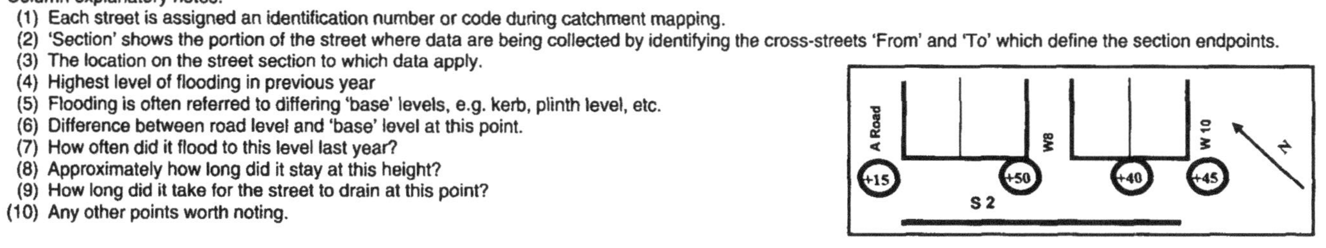

Direct observation during floods is much more helpful in seeing how the catchment as a whole performs than it can be for making specific performance measurements. Direct observation of pre-painted rulers on telephone poles or walls can provide useful comparisons between storms, and valuable additional data for detailed modelling studies of floods where computer models are in use.

Resident gauges

Resident gauges are simple rulers on telephone poles or walls, which are marked in 10 cm coloured bands (see Figure 6.1). Resident volunteers are supplied with notebooks in which they can record their observations of the levels reached by flood water, the time at which these levels were reached, and the duration of the flood (see Table 6.3).

Resident gauges have the advantage that they are cheap, they allow some estimation of the duration of flooding, and residents are much more likely than researchers to be around when a flood occurs. In addition, residents are distributed across the catchment, and can take simultaneous readings in a way that may be difficult for a small survey team. A system of resident gauging can be a useful way of involving the community in the study, but also runs the risk of raising expectations of relief from flooding which agencies may be unable to fulfil.

The system has the disadvantage that it depends upon the goodwill and diligence of resident volunteers. Residents have many other demands on their time, and may not be near the gauge at the peak of the flood. In addition, recording of floods at night is likely to be poorer than those which occur in broad daylight.

Our work in Indore achieved mixed results with resident gauges. In some cases, residents went to surprising lengths to record the hydrograph level of a major flood, but in most cases the data reflect great uncertainties about the start and finish of events. Very few residents managed to record the majority of events. The method is a good way to confirm answers to some very broad but important questions about duration; was the area flooded for only a couple of hours, or was there still significant depth a day later?

Chalk gauges

These consist of metre rulers coated with chalk, built into a small protective structure into which flood waters are free to enter (see Figure 6.2). As the flood rises, chalk washes off the portion of the ruler below the surface of the water. At the end of the storm, an approximation of the highest level of flood water can be read at the dividing line between the chalked and unchalked portion of the ruler. The metre stick is then coated again with chalk paste, ready to record the next storm's maximum flood level. The structure is needed to protect the gauge against children and vandalism.

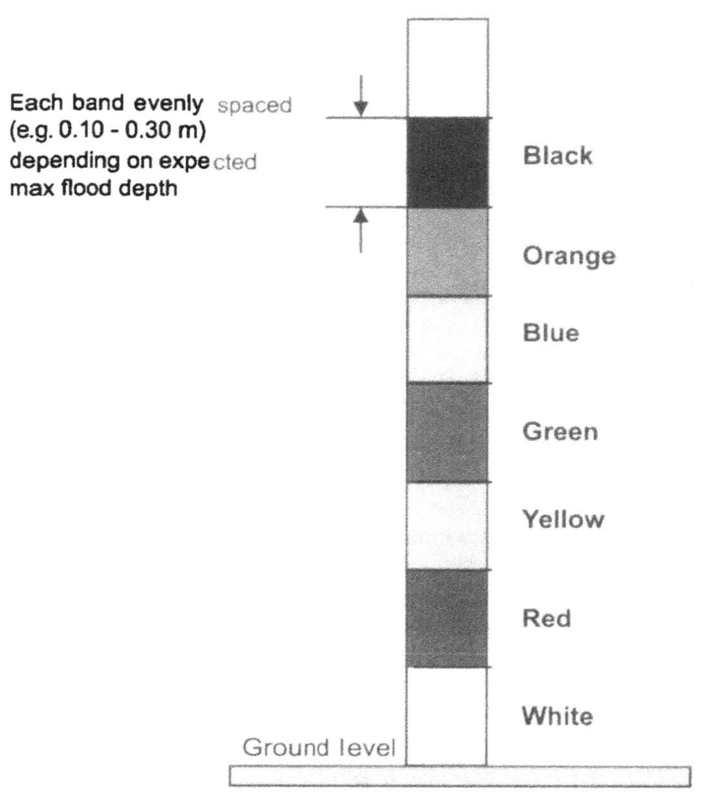

Figure 6.1 *Resident gauge for simple depth of flood measurement*

In Indore, we established a network of chalk gauges in each of two study catchments to record maximum flood levels. Chalk gauges have the advantages that they cost little, can be distributed across the catchment, and can record maximum flood levels at any time of day or night. In Indore, we found chalk gauges an invaluable check on the validity of computer models of drainage performance; our confidence in both the modelling and the validity of the measurements has been boosted by consistent agreement between the two. Chalk gauges have the disadvantage that they do not record the duration of flooding, and they require a regular process of reading and repainting after each storm. In addition, the gauges need to be carefully sited to provide useful information; it is generally best to site them in low areas where minor events can trigger flooding and thus record a level. Gauges that are too high will simply never record a reading.

Table 6.3 A page from a resident gauge notebook in Indore

Indore Drainage Evaluation Project

RESIDENT GAUGE OBSERVATIONS

Gauge No. 9 Sheet No. 5

Observer's Name & Address M.G. Ibrahim

567/S Nehru Nagar

दिनांक 19-8-94
DATE

वार Friday
DAY

समय 5.30 PM.
TIME

पानी किस निशान तक चढ़ा (✓ करें)
WATER ROSE TO WHICH MARK

नही चढ़ा DID NOT RISE	सफेद WHITE 1	पीला YELLOW 2 ✓	हरा GREEN 3	नीला BLUE 4	लाल RED 5

पानी उस निशान पर कितने समय ठहरा 2 Hours
HOW LONG WATER STOOD OVER THIS MARK

रोड पर से पूरा पानी उतरने में कितना समय लगा। 5 Hours
HOW LONG IT TOOK TO DRAIN COMPLETELY FROM THE ROAD SURFACE

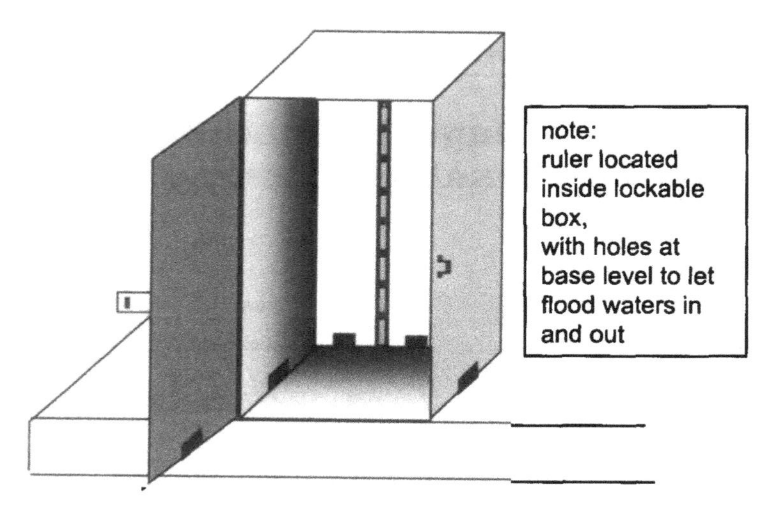

Figure 6.2 *Chalk gauge structure*

Electronic level gauging

The use of pressure transducers connected to data recording equipment (known as a 'logger') offers the most convenient form of data on flood depths. Like chalk gauges, this equipment can record data twenty-four hours a day; unlike chalk gauges, it can give a record of flood depth *over time*, so that durations can be easily computed. A typical depth log during a storm is shown in Figure 6.3. As the pressure sensor is located in a storm sewer, more than 1.4 m below the ground, no flooding took place on this occasion.

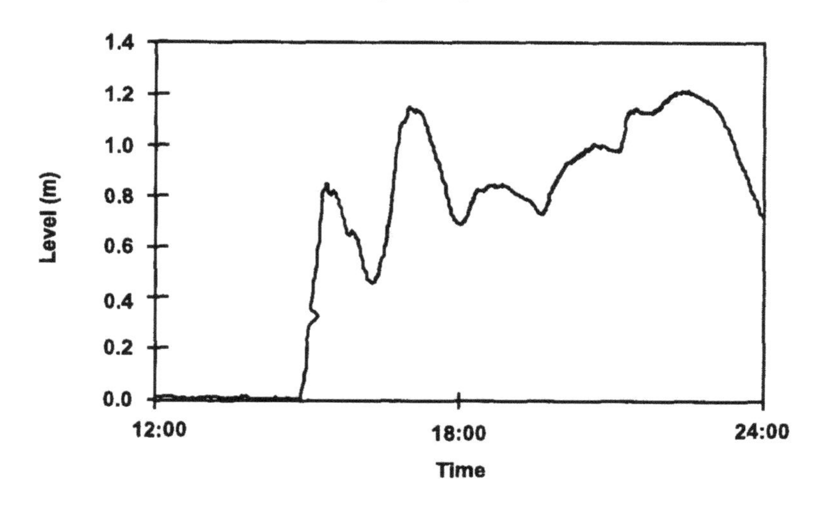

Figure 6.3 *Depth record from electronic pressure recording gauge*

Pressure transducers are often combined with electronic velocity gauges to permit discharge measurement in open channels; once the depth of flow is known, the section of flow can be calculated, which, multiplied by the velocity, yields the discharge. Thus a typical installation (Figures 6.4, 6.5) combining pressure and velocity sensors with recording equipment permits level and flow measurement within the drain, and may also offer flood level data, depending upon the level to which the water rises.

The major drawback of pressure transducers, velocity gauges and data loggers is their prohibitive expense. Like any other measuring device, they must also be checked for accuracy. This is relatively easy for depth, but much more difficult for velocity. In summary, these tools are indispensable for a detailed hydraulic modelling study, but are too expensive and complex for routine use in low-income communities.

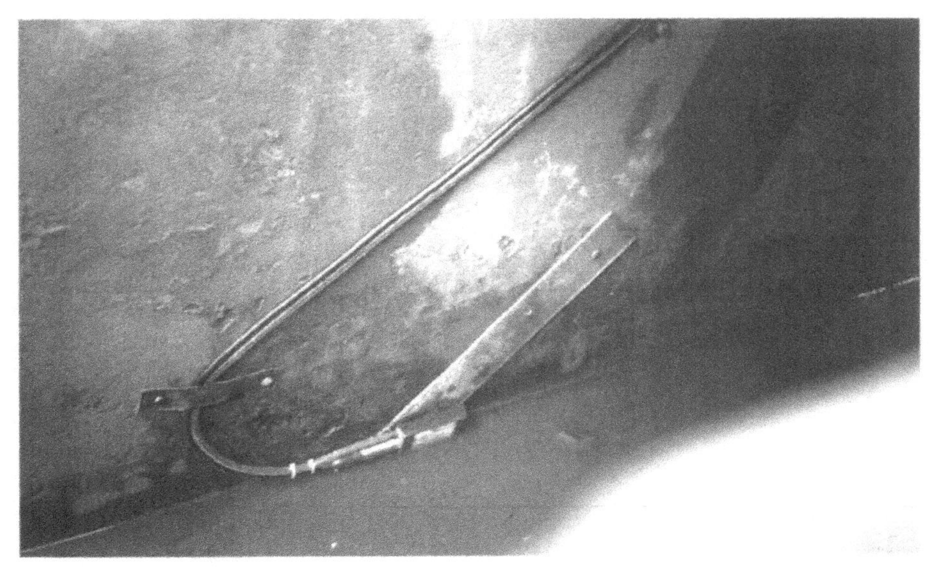

Figure 6.4 *Pressure and velocity sensor (photo: J. Parkinson)*

Summary

A systematic resident survey is probably all that is needed to assess the extent of flooding from a practical point of view. Other methods have been described in this chapter for consideration by those who may wish to do detailed studies for research or model calibration and verification.

Figure 6.5 *Above ground logging (data recording) equipment*

7 Flow estimation

As shown in Chapter 2, the volume and rate of runoff depend upon catchment area, ground cover type, and rainfall pattern. Very early in the study, it is helpful to have an approximate idea of the size of flows at different points in the system. This can be done by simply looking at catchment areas and indicative intensities, without considering cover type too closely. From the rational method of runoff estimation, outlined in Chapter 2, page 6,

$Q = kCiA$, where

$k = 2.78$ (for unit conversions) when

Q is in litres/second,
C is a dimensionless runoff coefficient,
i is rainfall intensity in mm/hr, and
A is catchment area in hectares.

This implies an absolute upper limit of flow/ha of about 2.8 l/second for each mm/hr of intensity. If a design rainfall intensity of 25 mm/hr is used to look at a sub-catchment of 4 ha, then an upper limit of design flow can be quickly estimated as $2.8 \times 4 \times 25 = 280$ l/sec. This is based on the simplifying assumption that the catchment is completely impervious.

Catchment area and cover type

The above type of quick computation can give an approximate idea of the relative magnitudes of flows at different points in the system. The effects of cover type must, however, be included before comparing flows with drainage capacities. The topographic and cover survey (Chapter 5) should allow estimates of both contributing areas and cover types at various critical points in the system. For the 4 ha subcatchment example, cover data might look something like those shown in Table 7.1. The 'effective area' is the equivalent area of impermeable area yielding the same runoff, and is simply the product of the runoff coefficient and the area. The last column shows the flow from the 25 mm/hr intensity assumption.

This table yields a much lower flow than the earlier estimate of 280 l/sec, and reflects the effect of cover type. As the urban community develops and open lots fill in with housing and pavement, the impervious area, and the runoff, increase. In the above example, we can estimate the effects of urbanization by assuming that as the open lots are developed, two-thirds are

http://dx.doi.org/10.3362/9781780446059.005

Table 7.1 Subcatchment details broken down by cover type

Cover Type	*Area (ha)*	*Runoff Coeff.*	*Effective Area (ha)*	*Flow (l/sec) (i = 25 mm/hr)*
Roof and paved road	1.2	1.00	1.20	83
Garden/green space	1.3	0.40	0.52	36
Open lots	1.5	0.50	0.75	52
Total	4.0		2.47	171

Table 7.2 Subcatchment details under increased urbanization

Cover Type	*Area (ha)*	*Runoff Coeff.*	*Effective Area (ha)*	*Flow (l/sec) (i = 25 mm/hr)*
Roof and paved road	2.2	1.00	2.20	153
Garden/green space	1.8	0.40	0.72	50
Open lots	0	0.50	0	0
Total	4.0		2.92	203

converted into housing and roads, while one-third become green space. Such analysis is helpful when considering plans for future developments, and assessing their impact upon runoff. Table 7.2 shows nearly a 25 per cent increase in runoff due to the effect of increased urbanization.

Rainfall intensity

The other great influence on flow, besides total area and type of cover, is the rainfall intensity. This is one of the most uncertain aspects of drainage evaluation. There are several approaches that can be used, depending upon both the data and staff resources available. These include:

- Use of a full set of intensity-duration-frequency (IDF) curves derived from locally available continuously recording rain gauges.
- Use of an approximate IDF curve derived from limited local data (e.g. hourly or daily rainfall data).
- Use of a single design intensity, e.g. 25 mm/hr (possibly varying with duration or catchment area), from a handbook.

Most hydrology and drainage textbooks (e.g. Chow *et al.*, 1987; Garg, 1994) describe the use of IDF curves. These curves show how the intensity of rainfall decreases with both duration and frequency. The maximum intensity of rainfall over 10 minutes is much higher than the maximum intensity of rainfall over an hour, for a given return period; the most intense hour often includes the most intense 10 minutes, along with 50 minutes of less intense rainfall, and this automatically brings down the average.

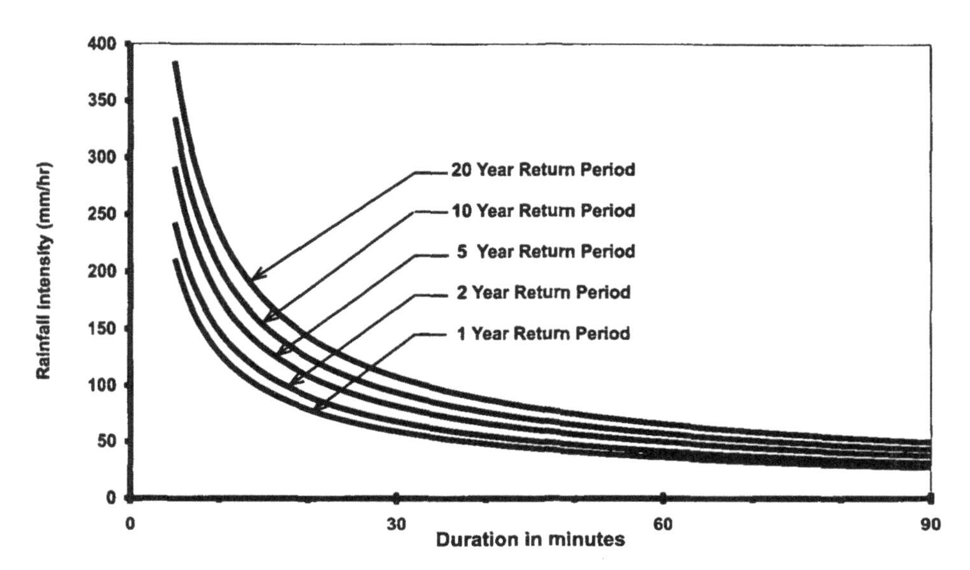

Figure 7.1 *Example of IDF curves (after Kothyari and Garde, 1992)*

Figure 7.1 shows some typical curves. These are derived from the relationships of Kothyari and Garde (1992) for central India, which are described in further detail in Annexe 7-A. It is important to understand that these curves do *not* describe the rainfall pattern of individual storms; the curve for the five-year return period may have a 10-minute intensity from one storm, and a one-hour intensity from a completely different storm in a different year. Similarly, they do *not* describe the actual rainfall intensity as it occurs over time; the first 10 minutes of any storm are very rarely the most intense, and a variety of techniques have been developed for simulating actual rainfall variation with time. Nevertheless, simple IDF curves are a rational and practical way to estimate design flows for comparison with system capacity. The remainder of this section shows how to combine IDF curves with catchment data to estimate flows. Annexe 7-A presents a variety of methods for deriving IDF curves from limited data.

Using IDF curves to estimate flows

Find a suitable set of IDF curves IDF curves may already be available for the catchment, or for a nearby site. Check with the weather service and local universities to determine what data are locally available, and in what form. If local curves have already been developed, these should obviously be used, unless there is reason to suspect them of inconsistency. (It is always worth a quick check that total accumulated rainfall increases with duration; if it doesn't, then something is wrong!) In many cases, however, IDF curves will not exist. In this case, you will need to develop your own, as shown in Annexe 7-A.

Select a return period for the evaluation Choosing a return period fixes the level of protection for which the system is being evaluated. In a drainage design, the chosen return period reflects some judgement about the relative costs of drainage works and flood damage; the ideal return period is chosen such that any additional investment in flood protection will cost more than the extra flood damage it prevents. In practice, performing such an analysis is too complicated, uncertain, and time consuming to be undertaken on a regular basis. Judgement, experience, and common practice are usually substituted for rigorous economics. Common criteria include the selection of a five-year return period for business districts, and a two-year return period for residential areas. There are strong arguments (Metcalf and Eddy and Engineering Science, 1966; Cairncross and Ouano, 1991; Watson-Hawksley, 1993) for using short return periods in developing countries where the money available for investment in drainage is very scarce. Unfortunately, development of IDF curves showing return periods of less than a year is not common practice. Most IDF curves are based on the largest event of each year, and these cannot easily be extrapolated below the 1-year level of return period; as noted earlier, the pattern of variation *within* years is likely to be quite different from the pattern of variation *between* years.

In practice, it may be worth evaluating systems at a range of return periods. Where time is limited, it is recommended to start with a very short return period of one year or less, as this will identify the most frequent shortcomings. As we have seen earlier, the variation of intensity with return period is slight compared to the variation of runoff with cover coefficient, so it is more important to put effort into assessing the surface cover.

Select a duration In the rational method of runoff estimation, the highest flow is generally predicted for rainfall of duration equal to the *time of concentration* t_c. The time of concentration is generally assumed to be the time it takes for water from the hydraulically most remote portion of the catchment to reach the point where flow is being estimated. If a duration longer than this is chosen, there will be no addition to tributary area, but the intensity will diminish, thus yielding a lower flow. If a shorter duration is chosen, more intense rainfall will be assumed, but the farthest reaches of the catchment will not have time to contribute to the flow. Where flows combine from catchments of different sizes and times of concentration, it may be necessary to check the times of concentration of the different subcatchments. This is because a subcatchment with a large area and a short time of concentration could yield more flow than the entire catchment with a longer time of concentration determined by a very small but distant subcatchment.

The time of concentration (t_c) has two components: the time it takes to enter the artificial drainage channel (*inlet time*, t_i), and the time it takes to travel through the drainage network to the point at which flow is being estimated (*time of travel*). The inlet time is often taken as 15 minutes (Tayler and Cotton, 1993), although values can vary between five and 20 minutes (WEF/

ASCE, 1992). The shorter time is appropriate for heavily paved and steep catchments; the longer time is appropriate for flatter, less paved catchments, especially where substantial ponding is evident. Travel time is based on estimates of velocity through the channel, which usually vary between 0.6 m/sec and 1.0 m/sec. Thus, including conversions from seconds to minutes, we have:

$$t_c = t_i + \left[L_{drain} / (V_{drain} \times 60 \, secs/min) \right]$$

where L_{drain} and V_{drain} refer to the length of drain (in metres) and velocity of flow in the drain (in metres/second) respectively.

Example: For the four hectare example, calculate the flows for differing return periods under current and urbanized conditions, using the Kothyari and Garde curves. Assume an inlet time of 15 minutes, and a drain length of 200 m, designed to flow at 1 m/sec.

$$t_c = t_i + \left[L_{drain} / (V_{drain} \times 60 \, secs/min) \right]$$

$$= 15 \text{ min} + [200 \text{ m}/(1 \text{ m/sec} \times 60 \text{ secs/min})] = 18 \text{ minutes.}$$

Read an intensity and compute the flow The IDF curves for central India, shown above, yield an intensity of 85 mm/hr for a duration of 18 minutes and a one-year return period, and 117 mm/hr for a five-year return period.

Using the previous section's computation of an 'effective area' of 2.47 ha under current development, the estimate of flow from the one-year storm is

$$Q = 2.78 \, CiA = 2.78 \times (2.47 \text{ ha}) \times 85 \text{ mm/hr} = 580 \text{ l/sec.}$$

It is useful to consider the effects of uncertainty in inlet time. If the inlet time is only 10 minutes, then t_c = 13 minutes, and intensity is approximately 110 mm/hr, yielding

$$Q = 2.78 \, CiA = 2.78 \times (2.47\text{m ha}) \times 110 \text{ mm/hr} = 755 \text{ l/sec.}$$

Alternatively, if the inlet time is 20 minutes, then t_c = 23 minutes, and intensity is approximately 70 mm/hr, and

$$Q = 2.78 \, CiA = 2.78 \times (2.47 \text{ ha}) \times 70 \text{ mm/hr} = 480 \text{ l/sec.}$$

This uncertainty is significant in the simplified modelling under consideration and will not be overcome by more precise techniques for development of IDF curves. This problem is reduced as the size of the catchment increases, and the inlet time plays a less significant role in fixing the time of concentration.

Limitations of IDF curves and simplified methods

As the last example shows, there are significant limitations to the precision of IDF curves and their use in evaluating and designing drainage systems. An

IDF curve does not represent a single real storm, but is synthesized from many different storms. It does not represent a time sequence of intensities, and must not be used directly for this purpose in evaluating the functioning of detention ponds and storage facilities. Finally, for small catchments, their use can be sensitive to relatively arbitrary definitions of 'inlet time'. Given these limitations, why bother with IDF curves? Isn't it necessary to simulate real rainfall to get a more precise understanding of how the system behaves in reality? Alternatively, if IDF curves are so inaccurate, why not just adopt a uniform intensity, such as 25 mm/hr?

IDF curves offer a meaningful compromise between the rigour of computer simulation of individual events and the adoption of a single rainfall intensity, regardless of climatological region. By reflecting the reduction in intensity with travel time, IDF curves help to keep a sense of 'proportion' in evaluating the system, help to identify bottlenecks in design, and help to reduce the risk of grossly over-designing one part while under-designing another through the use of a single design intensity. They are a very approximate tool, but offer a more realistic basis on which to assess the 'balance' within a system than any uniform intensity. In addition, once developed, they are a quick and easy way to get a feel for the match between storm flows and capacity. For the level of evaluation described in this manual, therefore, IDF curves represent the most appropriate, if approximate, basis for estimating rainfall intensity and computing flows to compare with capacity.

Annexe 7-A: How to develop IDF curves

THE METHOD USED to develop IDF curves depends principally upon the type of data available. Where data from continuously recording rain gauges are available, these should be used as they provide direct measurements of intensity. Unfortunately, such data are rare for smaller cities in developing countries. By contrast, airports, agricultural research stations, universities and a number of other sites regularly record daily totals. These data do not directly measure intensities, and so are not as directly useful for IDF development as those from continuous gauges. A number of approximate methods have, however, been developed to estimate the IDF curves from daily totals. This annexe therefore shows two distinct sets of methods for curve development, depending upon whether or not continuously recording gauge data are available.

Developing curves from continuous data

The procedures by which IDF curves can be developed are conceptually straightforward where good data are available. They are also, however, very tedious, and involve some arbitrary decisions. Good descriptions of the necessary procedures are available from Kibler (1982), and are summarized in Chow *et al.* (1987).

If one has N years worth of data, the procedures consist essentially of the following:

1. Define a way to separate events, e.g. 'at least 3 hours without rainfall'.
2. Identify a series of durations to be analysed, e.g. 5, 10, 20, 30, 45, 60, 90 minutes. (These may depend upon the shortest interval for which data are available.)
3. For each duration to be analysed, identify the largest total rainfall occurring over this duration within each event. If the duration to be analysed is longer than the length of the event, then use the total rainfall of the event as the event's depth at this duration.
4. Rank the rainfall totals (depths) for each duration.
5. Assign return periods for rankings. One of the most common formulae is: $T = (N+1)/m$, where T is the return period, N is the number of years of data, and m is the rank of the depth in question. (For example, consider 25 years of data. The biggest depth at a given duration has a rank of 1, and therefore a

return period of (25+1)/1 = 26 years. The thirteenth biggest depth has a rank of 13, and therefore a return period of (25 + 1)/13 or 2 years.)
6. Plot depths (mm) as a function of return period for each of the durations to be analysed. Fit a line or curve through the family of points for each of the durations. Semilog paper is probably best.
7. Read off depths from the line at each of the return periods of interest for each of the durations.
8. Divide these depths (mm) by their corresponding durations (hrs) to obtain intensities (mm/hr).
9. Plot a curve of intensity vs. duration for each return period of interest.

If the continuous data are recorded in digital form, then the above process is straightforward, and much or all of the work can be automated by PC or programmable calculator. If, as is more likely, the data are from chart recorders, then the most tedious part of the work will be the conversion of the chart readings to depths.

Developing curves from limited data

Here, the problem is conceptually much more difficult, although the proposed techniques probably involve less work than when raw data are available! Three approaches are presented below which allow approximate estimates of IDF curves given only limited data. This section illustrates the use of daily and hourly data from the National Meteorological Service of India for the city of Indore, from 1979 to 1992 to estimate IDF curves.

A regional approach

Kothyari and Garde (1992) developed curves for all of India, subdivided by individual hydrological regions. Their relation for India is:

$$i_t^T = k \frac{T^{0.20}}{t^{0.29}} (R_{24}^2)^{0.33}$$

where i_t^T = the rainfall intensity (in mm/hr) of duration t hours with return period of T years,

R_{24}^2 = the 24-hour rainfall (in mm) with a two-year return period, and
k = a constant varying by climatological region in India.

Kothyari and Garde offer the values for k as shown in Table 7-A.1 for different parts of India.

To use their method, find R_{24}^2 by examining the greatest daily rainfall recorded in each year from the available data. Relevant precipitation records from 1979 to 1992 for Indore are shown in Table 7-A.2 (1981 is omitted, as data for August were not available). These data may be plotted on a graph of daily precipitation vs. return period. If semilog paper is easily available, it's worth trying a plot, with the return period on the logarithmic scale, and the precipitation on the linear scale. This will yield a graph similar to Figure

Table 7-A.1 IDF curve coefficients for different regions of India

Geographical Region	*k value*
Northern India	8.0
Eastern India	9.1
Central India	7.7
Western India	8.3
Southern India	7.1

Table 7-A.2 Maximum daily precipitation, Indore, Madhya Pradesh (1979–1992)

Year	*Max. Daily Precipitation (mm)*	*Date*	*Rank m*	*Return Period (N + 1)/m = 14/m (yrs)*
1979	107.6	04/08/79	7	2.0
1980	91.2	06/06/80	9	1.6
1982	83.4	06/09/82	11	1.3
1983	86.2	18/07/83	10	1.4
1984	244.4	10/08/84	1	14.0
1985	135.6	08/10/85	5	2.8
1986	190.9	15/08/86	2	7.0
1987	139.1	08/07/87	4	3.5
1988	106.1	04/08/88	8	1.8
1989	54.4	29/06/89	12	1.2
1990	140.8	23/08/90	3	4.7
1991	108.5	10/06/91	6	2.3
1992	52.2	22/06/92	13	1.1

7-A.1. If semilog paper isn't available, then simply take the logs of the values of the return period, and plot the precipitation values against the log of the return period on ordinary graph paper.

In this case, a very good fit to the data was obtained. By using a statistical regression feature in a computer spreadsheet package, a least-squares line was calculated with the equation $P = 68.6 \ln (T) + 56.5$ (where P = maximum daily precipitation in mm, for a given return period of T years). A fit by eye would have been just as good, and a good estimate of the 2-year 24-hour rainfall could also be obtained by eye. The equation gives a 2-year 24-hour precipitation value of 104 mm.

Substituting this value into Kothyari and Garde's equation gives:

$$i_t^T = 7.7 \frac{T^{0.20}}{t^{0.29}} \left(R_{24}^2\right)^{0.33}$$

$$= 35.7 \frac{T^{0.20}}{t^{0.29}}$$

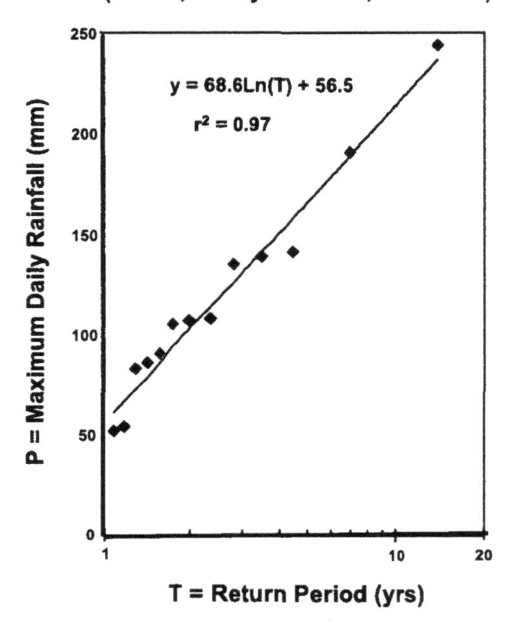

Figure 7-A.1 *Semilog plot of annual maximum daily rainfall vs. return period*

Table 7-A.3 Indore IDF relations from Kothyari and Garde's method

Duration	*Return Period (years)*		
(min)	*2*	*5*	*10*
10	146	176	201
15	110	131	151
20	89	107	123
25	76	91	105
30	67	81	92
40	55	66	75
50	47	56	64
60	41	49	57
75	35	42	48
90	31	37	42
120	25	30	34

Note that this approach has only required daily precipitation data. In their work, Kothyari and Garde have already studied the region-specific relationships between the 24-hour rainfall records and those for shorter durations. Their understanding of these relationships allows us to develop IDF curves

Table 7-A.4 Coefficients for IDF curves in various parts of the world

Location	*No. of stations*	*Range of durations*	*Average value of b*	*Range of values of n*
East Africa (a)	23	15 min–24 h	0.3	0.78–1.09
(b)	17	1 and 24 h	0.3	0.70–1.05
Ghana	14	0.2–24 h	0.6	0.86–1.03
Sri Lanka	19	3–24 h	0.5	0.74–0.96
Singapore	3	15 min–24 h	0.5	0.73–0.84
Sarawak	2	15 min–24 h	0.9	0.83–0.88
Sabah	3	15 min–24 h	0.9	0.86–0.99
Malay Peninsula				
East Coast	7	15 min–24 h	0.3	0.54–0.76
West Coast	11	15 min–24 h	0.5	0.86–1.05
Sumatra	1	30 min–24 h	0.3	0.95
Saudi Arabia	4	5 min–24 h	0.2	0.79–0.91
Barbados	2	5 min–24 h	0.2	0.73–0.78
Fiji	1	10 min–24 h	0.5	0.75
Hong Kong	1	5 min–24 h	1.0	0.91
UK	–	1 min–24 h	0.05	0.7
Libya	–	10 min–2 h	0.1	0.8
Dubai	–	15 min–2 h	0.3	1.0
Australia				
Brisbane	–	1 min–24 h	0.13	0.74
Sydney	–	1 min–24 h	0.08	0.64

(from Table 4.6, Watkins and Fiddes,1984, p. 42)

with less than ideal data. Similar work along these lines in other parts of the world may be extremely useful to drainage practitioners. Estimates for IDF curves for varying durations and return periods can now be computed, as in Table 7-A.3.

The approach of Watkins and Fiddes

The approach presented by Watkins and Fiddes (1984) is more broadly applicable than that of Kothyari and Garde, but still requires some knowledge of regional hydrology. Their approach is to find the values of a, b, and n in the following equation for intensity as a function of duration at any given return period T.

$$i_t^T = \frac{a}{(t+b)^n}$$

where i_t^T = the intensity (in mm/hr) of *t* hours of rainfall with return period of *T* years, and *a*, *b*, and *n* are coefficients. Watkins and Fiddes suggest the following:

1. *Start with a value of b from other studies in the region*; in this case, we can take $b = 0$ from the studies of Kothyari and Garde. For other regional values, see Table 7-A.4.
2. *Estimate the 'effective duration' of storms in the area*; the effective duration is the length of time over which 60 per cent of the day's rainfall occurs. This can be done by judgement, discussion with local experts, or a review of local storm data. As hourly data are available for Indore, a review of these allowed the construction of Figure 7-A.2, showing the distribution of t_{eff} among these storms.

Figure 7-A.2 shows the wide distribution of t_{eff} among the major storms in Indore, which makes the definition of a clear t_{eff} difficult. Forty per cent of the major storms, for example, showed 60 per cent of the day's precipitation occurring in less than three hours, while another 40 per cent of the storms revealed 60 per cent of the day's precipitation occurring over five hours or more. It is also not clear from hourly data whether a particular hour's rainfall is distributed uniformly throughout the hour, or whether it is concentrated in only a brief portion of the hour. An arithmetic average of the 'effective times' from the data yields a t_{eff} of about four hours, which is used in the rest of this section. However, the effect of choosing a different t_{eff} is also checked below.

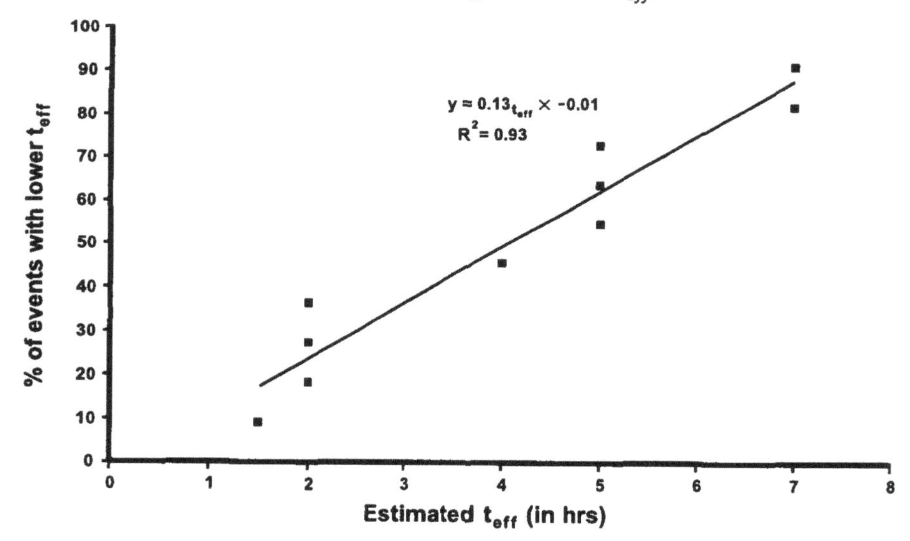

Figure 7-A.2 *Distribution of t_{eff} in major storms in Indore, 1979–1991*

3. *Given t_{eff}, compute a value for n.* This follows from algebra, given the value of i_{24} and the original equation used by Watkins and Fiddes:

$$n = \frac{\ln\left(14.4/t_{eff}\right)}{\ln\left(\frac{b+24}{b+t_{eff}}\right)}$$

In the case of Indore (and India in general, according to Kothyari and Garde's work), $b = 0$, so n simply becomes

$$n = \frac{\ln\left(14.4/t_{eff}\right)}{\ln\left(24/t_{eff}\right)}$$

4. *Compute a.* Finally, once n is known, a can be back-calculated from the 24-hour records from the equation

$$a^T = i^T_{24} \times (b+24)^n$$

5. *Compute IDF curve estimates.* As the calculations depend upon the value selected for t_{eff}, Table 7-A.5 shows the estimates of n, a, and intensity at 2, 5 and 10-year return periods using two t_{eff} values of two and four hours. The corresponding intensity-duration curves for the two-year return period are shown in Figure 7-A.3.

Table 7-A.5 Indore IDF computations using method of Watkins and Fiddes

Assumed t_{eff} (hrs)		4			2	
Resultant n		0.715			0.794	
Return Period (yrs)	2	5	10	2	5	10
24-hr Rainfall (mm)	104.0	166.7	214.2	104.0	166.7	214.2
i^T_{24} Intensity (mm/hr)	4.33	6.95	8.93	4.33	6.95	8.93
Estimated a	154	279	384	187	361	513
Duration (min)	Intensity (mm/hr)			Intensity (mm/hr)		
10	220	352	453	151	243	312
15	160	256	329	113	182	234
20	127	204	262	92.2	148	190
30	92.3	148	190	69.0	111	142
45	67.0	107	138	51.7	82.8	106.4
60	53.0	85.6	110	42.1	67.4	86.7
75	44.7	71.7	92.2	35.8	57.5	73.9
90	38.7	62.1	79.8	31.5	50.5	64.8
120	30.9	49.5	63.6	25.6	41.1	52.8

As Figure 7-A.3 shows, the IDF curves produced are fairly sensitive to the estimate chosen for t_{eff}. Apart from this, however, the method is broadly applicable wherever some hydrology studies have been made in the region (permitting an estimate of b). The advantage of the method lies in requiring only 24-hour rainfall data, and an estimate of t_{eff}.

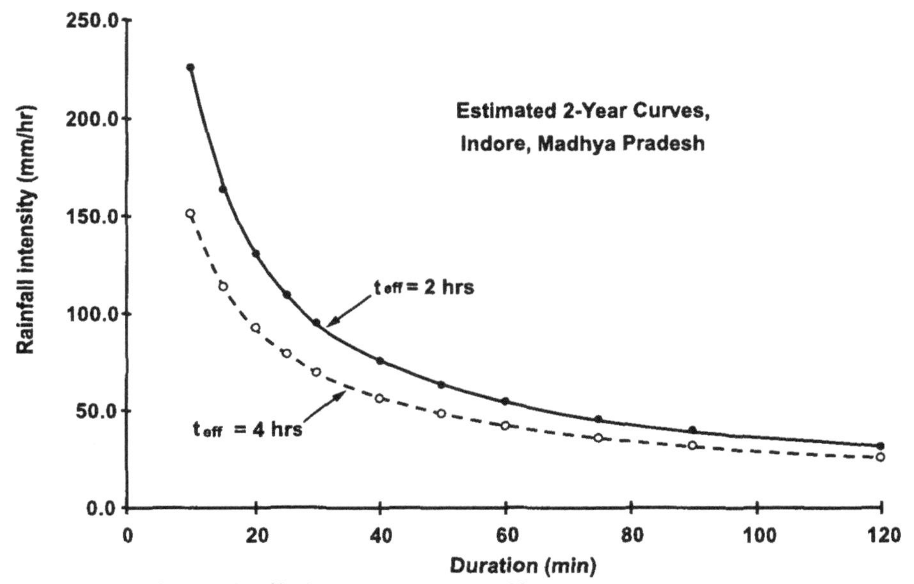

Figure 7-A.3 *Effect of differing t_{eff} values on Watkins and Fiddes IDF estimates*

Bell's method

Bell's method computes intensity as a function of duration and the one-hour precipitation for any given return period by the relation:

$$i_t^T = \frac{(0.54t^{0.25} - 0.50)\,P_{60}^T}{t}$$

where i_t^T = the rainfall intensity (in mm/hr) of return period T and duration t minutes, and P_{60}^T is the 60-minute precipitation of return period T. Bell's method was developed using North American catchments, but has been widely applied elsewhere for storms of less than two hours' duration.

Where hourly totals are available, Bell's method is straightforward to apply. An estimate is made of P_{60}^{10} or P_{60}^2. This can be done using a semilog plot as before, and as shown in Figure 7-A.4.

From the least-squares regression line, P_{60}^2 = 44.8 mm, and P_{60}^{10} = 67.8 mm. Substitution of either of these values into Bell's equation yields the corresponding IDF curve for the return period of interest.

Comparison of methods Figure 7-A.5 shows the IDF from all three methods for a two-year return period and Figure 7-A.6 shows the curves for the 10-year return period. The agreement for the two-year storm is striking, and may be partly due to chance. For the 10-year storm, the curves show more of the expected scatter. These methods are very approximate, but offer a better basis for evaluation than an arbitrary fixed value of intensity.

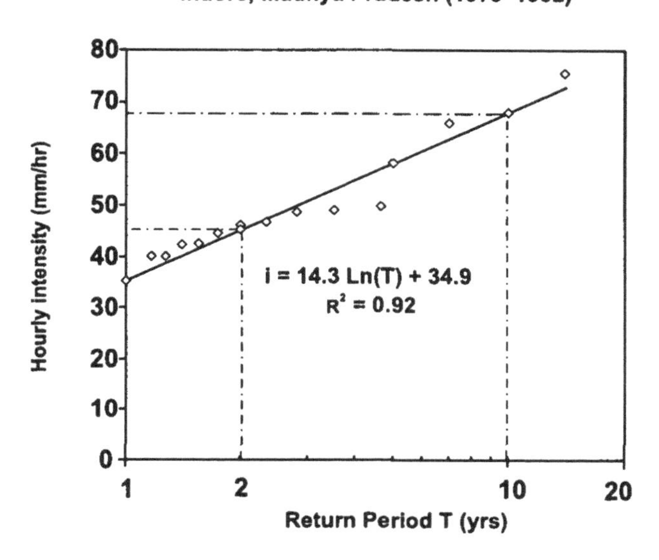

Figure 7-A.4 *Semilog plot of hourly rainfall intensities vs. return period*

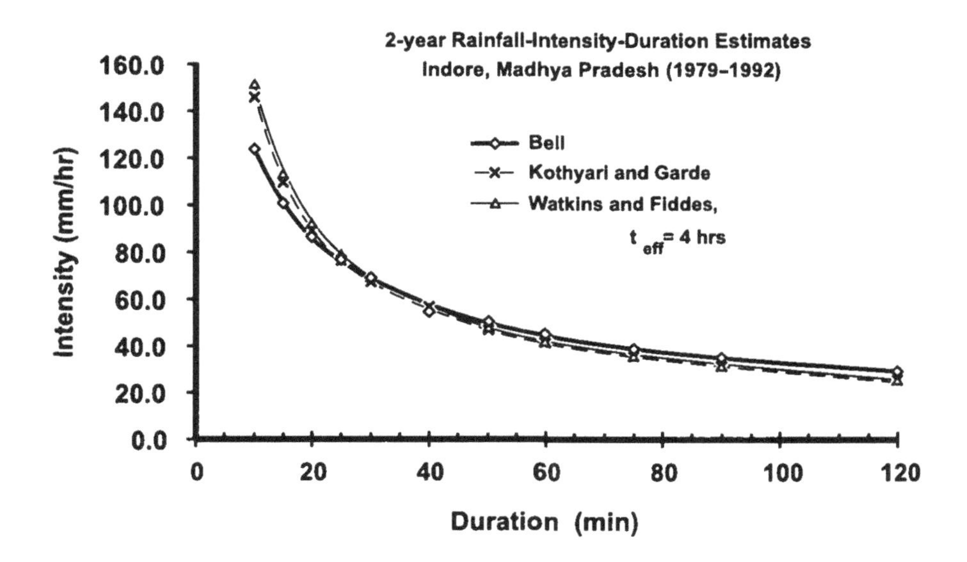

Figure 7-A.5 *Estimate of two-year IDF curves from three different methods*

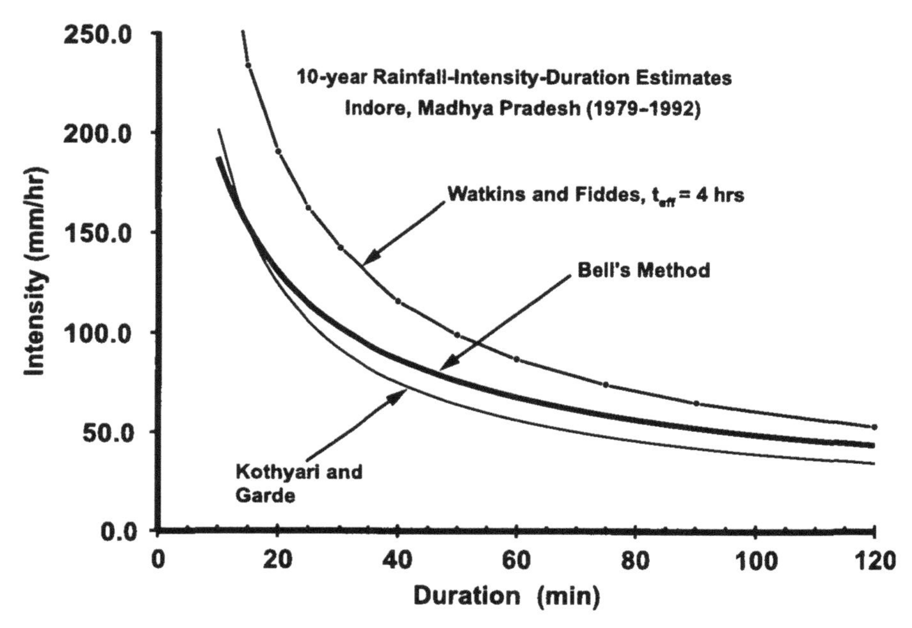

Figure 7-A.6 *Estimate of 10-year IDF curves by three different methods*

In general, the more local the data available for analysis, the better. Regional analyses, such as those of Kothyari and Garde, offer a useful compromise where few local data are available. If no local or regional data are available, but hourly data are, then Bell's method appears better than nothing, and offers reasonable agreement for Indore. Finally, if only daily precipitation data are available, but an 'effective rainfall period' can be identified in which 60 per cent of the rain falls, then one can try the approach of Watkins and Fiddes.

8 Assessing drainage capacity

THE LAST TWO chapters have looked at how to assess flooding which occurs in a catchment and the flows that provoke it. The rest of this manual looks at how to assess the other side of the coin: the drainage system intended to manage the runoff. This chapter shows how to estimate hydraulic capacity for comparison with flows estimated by the methods of Chapter 7. Chapters 9 and 10 will present aspects of the field measurements required for capacity analysis, while Chapter 11 describes field observation of drainage performance during storms.

Concepts of capacity

What do engineers mean when they refer to the 'capacity' of a drainage network? In general usage, the capacity is the largest flow that can pass through the network without causing flooding. There are a number of aspects to this definition that need to be thought about carefully:

- In the above definition, capacity appears to be defined precisely in terms of a single value of flow (e.g. 1.3 m^3/sec). In fact the capacity of a network is not a fixed single number, but will vary with circumstances. One obvious example is a drainage system that discharges to a river; the capacity is reduced when the river level submerges the outlet. Alternatively, the discharge capacity may increase with the hydraulic slope when the upstream water levels 'back up' above the crown of the pipe, but the outlet remains free. Drainage systems can also manage higher flows at the beginning of a storm (as empty channels and pipes fill) than they can later on when all storage is filled. Given these and other variations in hydraulic conditions, it becomes clear that no single number can define the maximum rate of runoff that may be managed.
- A drainage network is like a chain; it is only as strong as its weakest link. The capacity of a network can therefore be assessed only by determining the capacity of its individual components. This means looking not only at the capacity of the pipes or conduits, but also at the capacity of inlets.
- As earlier chapters have shown, it may be helpful to reconsider what is meant by 'flooding'. If it floods 1 cm deep for 10 minutes, have we really exceeded the system's capacity in a meaningful way? Does the major drainage system of flow along roads and streets really have no capacity?

http://dx.doi.org/10.3362/9781780446059.006

Or is it helpful to define different levels of capacity explicitly? For example, minor drainage system capacity is defined in terms of the discharge which causes any overflow, whereas major drainage system capacity can be defined by overtopping of house plinths or street kerbs.

- Capacity is estimated in different ways by different engineers and authorities. As just one example, capacity estimates will vary with the local authority's requirement for 'freeboard', defined as the minimum allowable distance between the highest predicted water level and the top of the channel. Freeboard is a safety factor, and is intended to make allowances for many areas of uncertainty and simplification in hydrology, construction, and hydraulics. Determination of a suitable freeboard standard is inherently uncertain, and different drainage and flood control authorities have varying freeboard requirements which imply different capacity estimates for the same physical system.
- The effect of solids and other blockage is rarely considered explicitly in capacity computations; it is usually assumed that the pipes are clean, and unblocked.

Three types of capacity estimation

There are three types of capacity worth considering in a drainage evaluation: design capacity, as-built capacity, and actual capacity. All three are illustrated in the next section. These capacities can be understood as follows:

Design capacity

- is based on design plans and drawings
- is straightforward to compute
- may usefully identify critical bottlenecks
- serves as an initial estimate of the upper limit of capacity for the existing system.

As-built capacity

- is based on field measurement of 'as-built' dimensions and levels
- is usually less than design capacity
- does not consider blockages or obstruction by solids
- serves as a more realistic upper limit of capacity without major construction changes.

Actual capacity

- makes explicit allowance for solids levels, and actual blockages
- requires significant field work and judgement
- is more approximate, but more meaningful, than other measures.

In approaching capacity evaluation, much depends upon the quality of information already available. The computation of design capacity makes sense only where good clear drawings are readily available. The other capacity estimates are always more meaningful, but if good drawings are available, a quick estimate of design capacity may help identify areas worth studying in greater detail. Where good drawings are *not* available, it makes sense to move directly to as-built and actual capacities.

Capacity estimates are essentially open channel hydraulic calculations, for which there are many textbooks (e.g., Chow, 1959; Henderson, 1966; Francis, 1975). A review of these books, and of the hydraulic software available for the solution of these problems (e.g. DoE/NWC, 1983; Huber and Dickinson, 1988), reveals the complexities of the subject. The hydraulics of storm drainage are particularly problematic because flows and levels change quickly with time; steady-state approximations usually underestimate the capacity of drainage systems as they ignore important storage effects.

Many hydraulic standards are also determined on the basis of local practice and judgement, so no universally acceptable criteria can be proposed. The guidelines offered below therefore suggest only a simplified approach, which will permit identification of the most severe capacity constraints.

In the capacity evaluation, there are two related numbers that are worth obtaining: discharge capacity and design head loss. For conduits of some length, the 'hydraulic capacity' as computed below is useful and can help to identify severely undersized stretches. However, such capacity must be kept in perspective; a pipe one metre long with no slope has 'no capacity' but this does not mean that water cannot flow in the pipe! For this reason, a more revealing number is the estimated head loss over the length of the conduit for a specific design flow. A table of head losses for each of the conduits and obstructions for a specified design flow can help concentrate attention where it may be most needed.

Design capacity For the assessment of design capacity without recourse to computer software:

- Assume the channel is clean.
- Assume uniform flow (i.e. assume depth does not vary, and water surface is parallel to channel bottom).
- Use a locally acceptable hydraulic equation. The Manning equation is strongly recommended for its simplicity:

$$Q = A \times V = A \times \left(\frac{1}{n} R^{2/3} S^{1/2} \right)$$

- where

  Q = discharge in m³/sec,

  V = velocity in m/sec,

  A = cross-sectional area of flow in m^2

  R = hydraulic radius (A divided by P) in m

S = slope of the bottom of the channel in m/m
n = the Manning roughness coefficient
P = the wetted perimeter, computed as the length of the section's perimeter in contact with the flow, in m

- In closed conduits, flowing full *or nearly full* include the top as part of wetted perimeter.

As-built capacity

- Use hydraulic criteria and analysis as above for design capacity.
- Difference lies only in that the data used for the computation are based on actual field measurement of conduit dimensions and levels.

Actual capacity

- Measure solids levels (see page 106).
- Treat solids as a solid physical barrier, and apply a suitably higher roughness value (e.g. Manning n of 0.030).
- Check for blockages (e.g. formwork left in manholes).
- Allow some estimate of head losses for these blockages.
- At points of low capacity, revisit, and identify the likely routes along which floodwaters would flow, and their consequences.

From these criteria, it is apparent that the first two 'capacities' are fairly routine hydraulic calculations with which engineers are generally familiar. Actual capacity evaluation involves more observation, more field data collection, and more judgement. Actual capacity estimates lack the mathematical 'precision' of conventional capacity evaluation, but are likely to be more meaningful. The insight gained from additional fieldwork will almost certainly highlight problems that conventional capacity estimation would overlook.

Examples of the three levels of analysis

Design capacity

Design drawings in the head office show a 500 m open concrete rectangular channel 0.6 m high by 1.0 m wide. The drawings show a uniform drop of 1 m over the 500 m length of the channel. Calculate the design capacity.

$A = 0.6 \times 1.0 = 0.6\ m^2$
P = wetted perimeter = 2×0.6 (sides) + 1×1.0 (bottom) = 2.2 m
$R = A/P = 0.27$ m
$n = 0.020$ (for rough concrete)
$S = 1/500 = 0.002$
$V = 1/n \times R^{2/3} \times S^{1/2} = (50) \times (0.27^{2/3}) \times (0.002^{1/2})$
$= 50 \times 0.418 \times 0.0447 = 0.93$ m/sec
$Q = A \times V = 0.6\ m^2 \times 0.93$ m/sec $= 0.56\ m^3$/sec.

As-built capacity

Field work determines that the actual slope is only 1/1000, and that encroachment by adjacent shops and homes has covered most of the drain. The section dimensions, observed at several points, are as in the design. What is the as-built capacity, i.e. the full uniform flow ignoring the effects of blockages and solids? Compare the capacities both including and excluding the top of the conduit.

A. Excluding the top of the conduit

As before,

$A = 0.6 \times 1.0 = 0.6\ m^2$

P = wetted perimeter = 2×0.6 (sides) + 1×1.0 (bottom) = 2.2 m (as before, ignoring the effect of the top of the conduit)

$R = A/P = 0.27$ m

$n = 0.020$ (for rough concrete)

Therefore the only effect on velocity and capacity is the reduction in slope.

$S = 1/1000 = 0.001.$

$V = 1/n \times R^{2/3} \times S^{1/2} = (50) \times (0.27^{2/3}) \times (0.001^{1/2})$

$= 50 \times 0.418 \times 0.0316 = 0.66$ m/sec

$Q = A \times V = 0.6\ m^2 \times 0.66$ m/sec $= 0.40\ m^3$/sec.

B. Including the top of the conduit

In this case, the wetted perimeter has changed significantly, as now the top is contributing to frictional resistance against the flow.

$A = 0.6 \times 1.0 = 0.6\ m^2$

P = wetted perimeter = 2×0.6 (sides) + 2×1.0 (top and bottom) = 3.2 m

$R = A/P = 0.188$ m

$n = 0.020$ (for rough concrete)

$S = 1/1000 = 0.001$

$V = 1/n \times R^{2/3} \times S^{1/2} = (50) \times (0.188^{2/3}) \times (0.001^{1/2})$

$= 50 \times 0.328 \times 0.0316 = 0.52$ m/sec

$Q = A \times V = 0.6\ m^2 \times 0.52$ m/sec $= 0.31\ m^3$/sec.

In this case, inclusion of the top of the conduit reduces the computed velocity and capacity by more than 20 per cent. When flow is near the top of the conduit, small waves or disturbances are likely to lead to contact between the flow and the conduit's top surface. Once this occurs, the top will drag on the flow, decrease its velocity, and increase its depth, thus ensuring that the flow stays in contact with the conduit top. For this reason, it is best not to count on the higher capacity of a partially full section, but rather to assume that the capacity is limited to that of full flow if there is any chance of the section 'temporarily' filling to the top.

Actual capacity

A solids survey is conducted at four points along the channel, which reveals a uniform solids depth of about 0.2 m of coarse material (brickbats, road metal,

etc.). Estimate the actual capacity at full flow, including the top as part of the wetted perimeter.
The cross-sectional area of flow has been reduced by one-third, so
$A = 0.4 \times 1.0 = 0.4\ m^2$.
P = wetted perimeter = 2 × 0.4 (sides) + 2 × 1.0 (bottom + top) = 2.8 m
$R = A/P = 0.143$ m
$n = 0.020$ (for rough concrete)
  = 0.030 (for the debris)
$S = 1/1000 = 0.001$
A simple average of n-value weighted by length may be used to allow for the extra roughness of the debris.
Thus,

$$n_{ave} = \frac{(n_{sides} \times l_{sides}) + (n_{top} \times l_{top}) + (n_{bottom} \times l_{bottom})}{l_{sides} + l_{top} + l_{bottom}}$$

In our case,

$$n_{ave} = \frac{(0.020 \times 2 \times 0.4) + (0.020 \times 1) + (0.030 \times 1)}{(0.40 \times 2) + 1.0 + 1.0} = 0.023$$

$$V = 1/n \times R^{2/3} \times S^{1/2} = (43) \times (0.143^{2/3}) \times (0.001^{1/2})$$
$$= 43 \times 0.273 \times 0.0316 = 0.37 \text{ m/sec}$$
$$Q = A \times V = 0.4\ m^2 \times 0.37 \text{ m/sec} = 0.15\ m^3/\text{sec}.$$

Note that in this example, the *actual capacity* is estimated as one half of the *as-built capacity*, and slightly more than a quarter of the *design capacity*! This illustrates the dangers of relying entirely upon design drawings, and the need for field studies to assess capacities.

Drainage network surveys

Good data collection is essential if you are to move beyond simple design capacities. Chapter 9 describes important aspects of the structural survey (to determine as-built capacity) while Chapter 10 describes maintenance surveys (to determine actual capacity). As in the catchment analyses described in Chapter 5, there is no alternative to good topographical survey information.

Good information requires an effective communication of requirements as much as a disciplined and competent survey team. For example, capacity estimation requires accurate invert levels and section dimensions. To provide this information, the survey team must learn to pay attention to such details as the difference between the bottom of the pipe and the bottom of the manhole. It is the pipe or channel sections which control the hydraulics, and where manholes are deeper than the corresponding pipes, only the conduit depths matter. These differences may be obvious to the engineer, but there is no reason to assume they are understood by the survey team, and the data requirements must be made explicit. Sample survey sheets are shown in Chapters 9 and 10, but well-designed survey sheets are no substitute for field work by the engineer with the survey team.

Level of network analysis

No matter how small the catchment being studied, some simplification will be required. For example, detailed study of flow in gutters upstream of storm drain inlets is undertaken only as a research activity; simplified models of overland flow such as 'inlet time' are used to consider these effects. In hydraulic analysis of any system, decisions must always be made about how much detail to include. The degree of simplification required depends a great deal upon the computational tools being used. Some believe that if computers are being used, it makes sense to include as much detail as possible. Others believe that perspective can be lost by including all sub-systems, as the mountains of data soon overwhelm our ability to understand them, or, for that matter, to check their correct entry into the computer. Reasonable limits and approaches to simplification are therefore suggested below.

If the analysis is being done by hand calculation (e.g. calculators), it is best to start with no more than 20 or 30 distinct hydraulic conduits. Some of these may need more detailed study later on, and a more detailed analysis might then be done separately. At the beginning, however, the need is to find the obvious weak links in the chain. Division of the network into individual conduits should therefore be based on both:

- hydraulic changes (e.g. significant changes in channel section or slope) and
- hydrologic changes (e.g. where new catchments enter the system).

If the system is well designed, these two should correspond, but this may not always be the case. Sectioning of the network should therefore take place after the definition of subcatchments, to simplify the comparison of runoff estimates with observed capacities. It is not realistic, with computation by calculator, to model each small change in channel section, slope, or flow; it will be too much work, and the engineer will spend too much time on small problems. Even where computers are used, there is a danger that too much effort can go into describing all the details without enough attention to what is going on overall.

In general, one of the best ways to simplify the work is to ignore the hydraulic analysis of the most upstream elements (while not ignoring their contribution to tributary area!). This is because downstream bottlenecks will affect the entire network, whereas upstream problems are local. If field work shows these upstream areas flood frequently, they may then be analysed separately, without altering the downstream analysis significantly. By contrast, simplifications of the main network may overlook bottlenecks which can affect the entire system.

Annexe 8-A: Using software for drainage analysis and design

Introduction

In many cities, computers have revolutionized the analysis of drainage problems and solutions. The development of drainage software has made the hydrologic and hydraulic analysis of complex networks a great deal easier. This manual draws extensively from modelling results using the high-quality packages developed by Wallingford Software Ltd. It may therefore seem surprising that this manual does not give much attention to computer modelling as a way of assessing performance. There are three main reasons:

- The wide and rapidly changing variety of available software. The best sources of information on specific aspects of drainage software are the suppliers themselves, and most will provide a good deal of information on request *before* you commit to buying a package. Given the availability of package-specific information from suppliers, there is little point in trying to cover the same ground superficially in a set of broad guidelines such as these.
- There is an extensive literature on drainage software. Books, academic journals and conferences regularly report on advances in this field; good discussions on modelling may be found in many current references (e.g. Chow *et al.*, 1987; WEF/ASCE, 1992). Most new publications on drainage are prepared in the industrialized world, with an assumption that computers are available and appropriate; there is little guidance for staff without access to such resources.
- A strong belief in the need for field work and common sense. Computer software is *not* essential to the analysis or understanding of every drainage problem, and this manual has attempted to point out how much insight can be gained without use of sophisticated software. Unfortunately, so much software is so easy to use *without* proper field work that the pitfalls of their improper use (summarized below) can often overcome the advantages. Skilful use of computer software can nevertheless be an invaluable extension to field work and common sense.

This annexe starts with a broad review of the types of software available, the basic assumptions built into the packages, and what they may or may not be expected to do. Some factors to consider in choosing between different packages are presented. The strengths and pitfalls of computer packages are then summarized.

Classification and characteristics of software

There is a wide variety of software on the market, offering a wide range of analytical powers. This variety can be classified in a variety of ways, but the most common distinctions are described below.

Simulation and design packages

Most packages are *simulation* software, that is, they simulate the flows and hydraulic performance of a specified system under a specified set of conditions. Simulation packages require the user to specify both the drainage system to be analysed (e.g. pipe dimensions, levels, and layout) and the hydrologic conditions (e.g. rainfall pattern, or specified design storm conditions). Simulation packages are often used to compare the performance of a range of designs; the use is often iterative, as the designer changes various components that limit the capacity of the system.

Design packages help the user to select appropriate dimensions and levels of new components of a drainage system more directly. Such software either includes internal design criteria, or asks the user to specify design criteria for minimum and maximum velocities, minimum and maximum slopes, etc. The user must also specify data on the site's topography, and the direction and length of drainage conduits.

Design packages have an intuitive appeal, as they appear to answer the engineer's design question more directly than do simulation packages. In practice, their use is often limited by the existence of drainage components that are *not* going to be replaced, the performance of which must be simulated. In addition, the hydraulic analysis used by design packages is often far simpler than that used by simulation software, which means that the design software may not give a very good idea of what will actually happen under storm conditions.

Many packages combine features of both types of analysis. Iterative design through simulation may appear to be a waste of time, but in practice the approach offers greater flexibility for the development of alternative solutions.

Level of hydraulic analysis

There are three basic levels of hydraulic analysis used in drainage software, with variations in their implementation between packages.

Uniform, steady flow The simplest and most common form of analysis assumes (a) that flow does not vary over time, and (b) that the depth of flow in a conduit is constant along its length. Such an analysis computes capacities of conduits in much the same way as shown in Chapter 8. It cannot analyse flow where there is no slope or where there is a backfall, i.e. where the downstream invert of the pipe is higher than the upstream invert. It can, however, serve as a quick check on capacities throughout the network, and can be easily set up on a computer spreadsheet without purchasing additional software.

Gradually varied, steady flow This analysis assumes steady flow, but allows the depth of flow to vary along the length of the conduit. This is a major improvement over uniform flow modelling, as it allows consideration of backwater effects, and the performance of components with flat slopes or backfalls. The computation involved in this analysis is much more complex than that for uniform flow models, and is not recommended for the amateur programmer!

While such packages offer an improvement over uniform flow, the assumption of steady flow is still not very realistic for most drainage systems, as rainfall, runoff, pipe flows, and levels vary quite rapidly during the course of a storm. Storage effects, both on the catchment surface and in the drainage conduits are important, as they alter the magnitude of the peak flows experienced by different components of the network.

Dynamic simulation packages These packages are the most realistic and the most complex. They allow simulation of the variation in water levels and flows over time, thus permitting analysis of important storage effects in the network. There are a number of ways to model flood waves travelling up and down the drainage network, including kinematic waves, diffusion waves, and solution of the full St Venant equations of momentum and continuity. The modelling results presented in this manual were obtained using the rigorous analysis of the Wallingford software packages which solve the full St Venant equations. The interested reader is referred to Chow *et al.* (1987) for further details on the various forms of dynamic hydraulic analysis.

While dynamic packages are extremely powerful, they also require great care and effort in their use, and enormous quantities of high quality data. Any potential user of a dynamic model needs to think long and hard about the availability of data on the variation of flow and water level over time; without such data, the validity of the model cannot be assured, and the extra cost and effort may be wasted.

Hydrologic analysis

A wide variety of hydrologic models exist to predict runoff from rainfall. The rational method, described in Chapter 7, is the simplest and most common basis for the estimation of runoff from rainfall. The method takes runoff to be a fixed proportion of a specified intensity of rainfall; this intensity is either specified as a design criterion (e.g. '25 mm/hr'), or else is determined from intensity-duration-frequency (IDF) curves. In the latter case, the user specifies the appropriate frequency of the design event, and the package computes the critical rainfall duration (equal to the time of concentration) at each point where flow is to be estimated. Given both the duration and the frequency, intensity is computed from the IDF curves, and then converted to runoff by the rational formula.

The rational method requires only a specified rainfall intensity, or the use of IDF curves, and does not require time-varying rainfall data. This simplicity comes at a cost; the method cannot consider storage effects in the catchment or drainage network, and the results are practically impossible to validate.

More realistic modelling of relations between rainfall and runoff is possible using the data from recording rain gauges as input. Such models no longer use the statistical averages of IDF curves, but instead model the conversion of rainfall to runoff during an individual event. A wide variety of such hydrologic models exist, including the Wallingford Procedure (DoE/NWC, 1983), the SWMM package (Huber and Dickinson, 1988), the modified SCS procedure, and various modifications of the rational method; many packages allow analysis by more than one approach.

While much excellent work has been done to develop and test these hydrologic models, the most critical issue from the user's point is the availability of catchment, rainfall, and runoff data with which to validate the model. Where such data are unavailable, the value of sophisticated hydrologic models is limited.

User interface

So far, drainage software has been considered in terms of what it computes. The 'user interface' determines the ways in which users interact with the software, data and instructions are entered, and results are displayed. A good interface simplifies data entry and verification, and presents results in an intuitively understandable way on-screen as well as in complete well-structured text reports. There is often a major tradeoff between a good user interface and cost, as user-friendly software costs more to program.

The requirements for user interface vary with the complexity of the model. Dynamic models require a better interface than steady flow models, as the results reflect variation with time; users will want to see graphs of water level and flow as a function of time at a variety of points in the network. A dynamic model that simply prints large tables of numbers is not going to be very helpful in making decisions.

Geographic information systems (GIS) link maps with database management systems. This allows users to inspect data associated with points, lines, or areas on a map. In practice, this can mean that a user can easily plot a profile of a drain, or can find the depth and condition of a manhole, through selection of the relevant components on the screen with a 'mouse'. GIS systems are complex but potentially valuable tools for urban management, as a variety of utilities can be included in the same maps and files. Many packages incorporate various GIS features, and where GIS is being used elsewhere in the municipality, system compatibility is an important consideration in selecting a drainage package.

Choosing between packages

Where much drainage analysis and design is going to be done, computers and software packages can be very worthwhile investments. This section summarizes some of the most relevant considerations in selecting a package.

Purpose

What kind of work is the software intended to do? If you want a quick overall review of an existing system to identify (or confirm) suspected bottlenecks, of the sort described in Chapter 8, then simple capacity models based on the rational method and steady uniform flow may be appropriate. This will *not* do a very accurate job of predicting performance, but can help to focus attention on capacity constraints. Such a model can be set up (with time and care!) on a simple spreadsheet package, without purchase of specific drainage software. (On the other hand, commercial packages may offer a variety of features for data entry and report writing that you wish to use, and cannot easily develop on your own.)

If the system is a large one, and the expected capital expenditures are also large, then more sophisticated models are appropriate; the time, effort, and expense of data collection and computer analysis may well be justified by the savings from developing more appropriate analyses and solutions. Similarly, academic research on drainage performance and improvement requires use of dynamic models, as only these sophisticated packages can actually predict the variation of level and flow with time. In either case, the cost and use of the model should be viewed as the 'tip of the iceberg'; the most pressing resource demands are for good data with which to build and validate the model.

Data requirements

The more closely a package describes physical performance, the more data are required to develop, calibrate, and verify the model. The natural desire to simulate the physical world 'as closely as possible' must be balanced against the availability of data with which to build the model, and the capacity to gather further data. If data from continuously-recording rain gauges, flow monitors and depth gauges are unavailable, there is little point in investing in dynamic hydraulic models. The use of computer packages does not eliminate or reduce the need for field studies described in the rest of this manual; indeed, all of the techniques shown were developed to help build and validate computer models of the two catchments considered.

Hardware and software requirements

Software suppliers will usually identify the minimum hardware and software requirements for their package in terms of processor, memory, hard-disk space, and operating system. You should treat these specifications with some scepticism, and question the supplier closely on this point; in some cases, the software may technically be capable of running on your system, but at such a slow speed as to be impractical. It may pay off to ask for a demonstration of the software on your hardware and software so you can observe performance for yourself.

Why invest in a software package?

- *Fast computation means more alternatives can be assessed.* Analysis of drainage systems is much quicker by computer than by hand, *once your model is set up*, permitting consideration of a wide range of alternatives.
- *Models often provide a useful framework and motivation for data collection.* Several modellers have found that the need to check data for a model has uncovered surprising facts about the system they are assessing. Strange model results which do not conform to experience have also prompted useful field checks that revealed errors in assumed values.
- *Careful model use can provide physical insight* into the nature of drainage problems that is otherwise difficult to obtain.

What are the pitfalls to avoid?

- *Uncritical acceptance of results.* Computer packages are, at best, only as valid as the data they use. The quality of the data must be checked carefully, both *before* data entry (to ensure that what is entered accurately reflects field conditions) and *after* data entry (to ensure the data were correctly entered). Extra care at the early stages of the study is critical; if data about the existing system are found to be erroneous half-way through the analyses, many computer runs may have to be repeated.
- *Unanswered questions may still be important.* Computer packages are designed to answer some questions and not others. Users need to be careful not to ignore important questions (e.g. 'What happens when it floods?') just because the computer package may not address it.
- *Reduced time in the field.* Wise use of a computer package should *increase*, not decrease, field work and data collection. Yet the appeal of high technology, computer simulation, and rapid results may tempt good staff to spend too much time behind a desk, and not enough time in the field. A mathematician wrote that 'the purpose of computation is insight, not numbers', yet drainage computation without field work may too easily produce the exact opposite.

9 Drainage network structural survey

As noted in the last chapter, it is helpful to separate 'as-built' capacity limits (which are built into the structure of the network) from other capacity problems caused by temporary blockages or solids deposits. These different constraints need different solutions to improve performance. In reality, the distinction is not always so clear; if a drain is laid on a very flat slope, solids can build up rapidly even with frequent cleaning. This may thus appear to be a 'maintenance' problem, when it is actually caused by poor design, poor construction, or both.

This chapter describes survey techniques to check the structural or 'fixed' conditions which limit as-built capacity. The next chapter will focus on measurement of maintenance and solids that further reduce capacity. One way in which the two types of surveys differ is that the formal structural survey should need to be done only once, with perhaps multiple visits to inspect complicated or important parts of the system. (Frequent informal 'surveys' of the drainage network, however, are essential in getting a sense of how the catchment works, as described in Chapter 5.) The maintenance surveys are worth repeating several times, to see how solids and debris change over time. In practice, however, the structural and maintenance evaluations will overlap, as both types of problem can become apparent during the same survey.

Conduit measurements

Dimensions

The dimensions of cross-sectional area and length of the various conduits in the system are critical data for capacity estimation, and can be measured simply with a tape measure. Field workers must understand exactly what is to be measured; the 'width' of a trapezoidal channel, for example, could mean many things. (It is usually best to measure both the bedwidth and the top-width; in an irregular section it is good to measure width at half the section height as well). Figure 9.1 shows some common definitions of terms.

In hydraulic terms, the inner dimensions are the important ones. No level is required to gather dimensions, but measurements of 'relative elevations' of soffits or inverts must be made with respect to a fixed point for which the level may be easily found. Field workers sometimes prefer to measure the outside dimensions of a pipe, as these are often more accessible. If this is done, the pipe thickness must be measured. Soffit elevations may also be easier to

http://dx.doi.org/10.3362/9781780446059.007

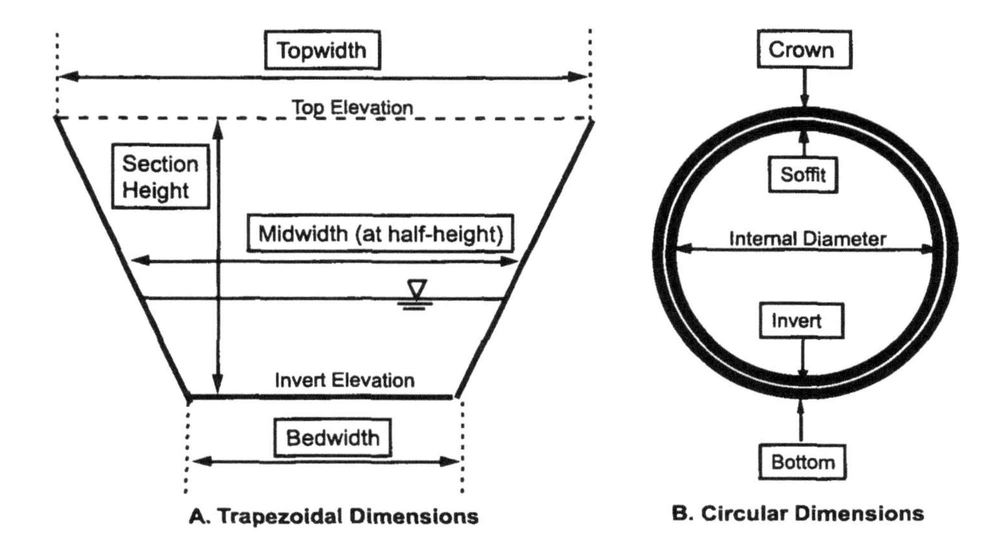

Figure 9.1 *Main conduit dimensions and elevations*

obtain than invert elevations, particularly when there is flow and sediment in the pipe. Provided the pipe is round or the section height is accurately known, the invert elevations can be worked out from the soffits. It may also be difficult to measure bedwidth, and extrapolation from topwidth and midwidth may be appropriate where the section is regular.

Levels

Levels are critical in determining capacities. The important levels are those of the conduits themselves (e.g. pipe soffits and inverts), and not those of the junctions connecting conduits. Surveys therefore involve measuring dimensions such as 'depth to pipe invert' from the frame (see Figure 9.2). Frames are not always level, so it is also worth marking the point on the frame from which the measurements were made. Once the frame levels are available, conduit levels can be established. Thus, if a frame elevation is 101.23 m and the depth to a pipe invert is 1.90 metres, the pipe invert elevation may be computed simply by subtraction as equal to 99.33 m.

Tables 9.1 through 9.4 show some survey sheets and forms used to establish dimensions, elevations and as-built capacities. Closed conduit systems are particularly troublesome because surveyors can observe the conduits only at manholes. Such a survey is much easier if a consistent naming or numbering system has been developed for manholes and pipes in the network as part of the original catchment mapping. This will evolve as more data become available (e.g. buried manholes or unknown connections emerge from the survey) and great efforts must be made to ensure that consistent numbering is used at different stages of the work. If new numbers need to be added, or old ones need to be dropped, the essential point is *not to change the*

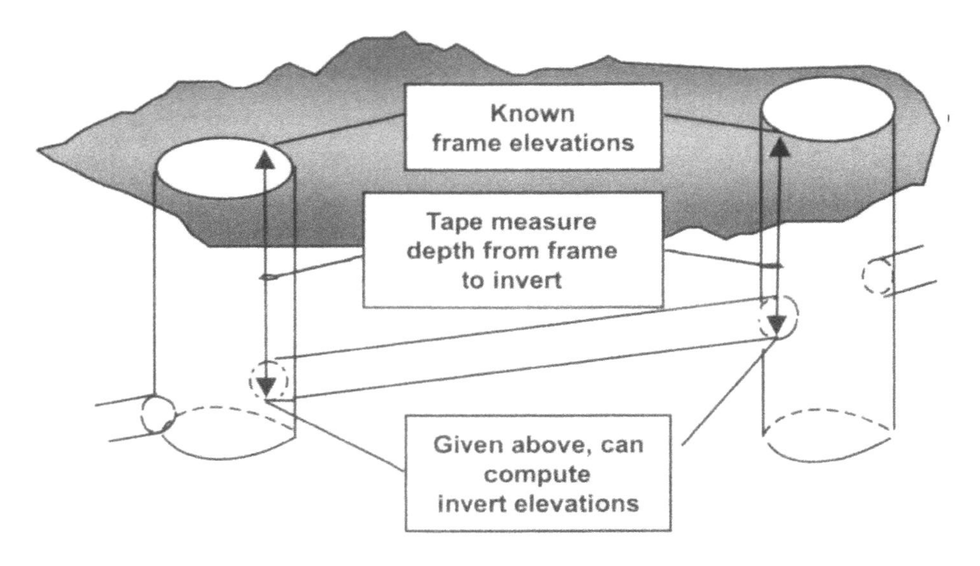

Figure 9.2 *Measurements required to establish invert elevation*

Table 9.1 Storm drain manhole survey sheet

Surveyor:
Date:

Manhole Ref. No.: CR2N4
Type: Junction only; no direct inflow.
Lid in place?: Yes
Structural condition of manhole:
Internal: Sound.
External: Chipping of frame.

(**Note**: Use paint to mark frame points from which depths are measured. These points can then be levelled. Do not mark covers, as these can be moved, rotated, and switched between manholes. Instead, paint some fixed part of the structure from which measurements are made.)

Pipe ID	*From/To*	*Size*	*Soffit Depth from Frame*	*Remarks*
Outlet	To CR2N5 (outlet)	450	1.2	Standing water 10 cm above invert
Inlet 1	From CR2N3	300	1.2	Dry
Inlet 2	street inlet	300	1.0	1/2 blocked with solids
Inlet 3	From R6N1	300	1.3	Standing water

(**Note:** It is best if a manhole identification system can be worked out in advance of the survey. However, the survey may well uncover pipes and manholes that have been forgotten, for which you have no identification. Identification systems can be developed afterwards, but *only* if each manhole's data are accompanied by a clear sketch. This is why the sketch should show some means of orienting the location of recorded data. As shown in the sketch above, it is helpful to use paint to mark frame points from which depths are measured.)

Remarks: Some blockage or backfall in outlet or downstream, as standing water observed.

Table 9.2 Table of pipe invert levels worked out from manhole survey sheets

		Outlet				Inlet 1				Inlet 2			
Field MH ID	*MH frame Elevation (from topo survey)*	*Pipe ID*	*Depth to Soffit*	*Pipe Size*	*Invert*	*Pipe ID*	*Depth to Soffit*	*Pipe Size*	*Invert*	*Pipe ID*	*Depth to Soffit*	*Pipe Size*	*Invert*
CR3N1	**544.38**	CR3N1.1	0.89	0.30	**543.19**	Extern 1	0.87	0.30	**543.21**				
CR3N2	**544.17**	CR3N2.1	0.86	0.30	**543.01**	CR3N1.1	0.84	0.30	**543.03**				
CR3N3	**544.08**	CR3N3.1	0.79	0.30	**542.99**	CR3N2.1	0.78	0.30	**543.00**				
CR3N4	**544.01**	CR3N4.1	0.76	0.30	**542.95**	CR3N3.1	0.72	0.30	**542.99**				
CR3N5	**543.94**	CR3N5.1	0.81	0.30	**542.83**	CR3N4.1	0.81	0.30	**542.83**				
CR3N7	**543.97**	CR3N7.1	0.84	0.30	**542.83**	CR3N5.1	0.86	0.30	**542.81**	house	0.71	0.10	**543.16**
CR3N8	**543.82**	CR3N8.1	0.63	0.30	**542.89**	CR3N7.1	0.64	0.30	**542.88**	house	0.35	0.10	**543.37**
CR3N9	**543.72**	CR3N9.2	1.04	0.30	**542.38**	CR3N8.1	1.04	0.30	**542.38**				
	543.72	CR3N9.1	1.13	0.45	**542.14**	This second outlet from CR3N9 goes to CR3S8							
CR3N10	543.65	CR3N10.1	1.01	0.30	**542.34**	CR3N9.2	1.00	0.30	**542.35**				
CR3N11	**543.53**	CR3N11.1	1.66	0.30	**541.57**	CR3N10.1	1.61	0.30	**541.62**				
CR3N12	**543.45**	CR3N12.1	1.54	0.30	**541.61**	CR3N11.1	1.56	0.30	**541.59**				
CR3N13	**543.65**	CR3N13.1	1.77	0.30	**541.58**	CR3N12.1	1.75	0.30	**541.60**	house	1.74	0.10	**541.81**
CR3N14	**543.46**	CR3N14.1	1.60	0.30	**541.56**	CR3N13.1	1.59	0.30	**541.57**	house	0.51	0.10	**542.85**

Table 9.3 Table of pipe capacities from inverts and survey data

Pipe ID[1]	*Down MH*	*Length (m)*	*Cum. Length (m)*	*Diam. (m)*	*Up Invert (m)*	*Down Invert (m)*	*Slope (m/m)*	*Capacity*[2] *(l/sec)*	*Remarks*
CR3N1.1	CR3N2	35	453	0.30	543.19	543.03	0.0046	57	
CR3N2.1	CR3N3	14	439	0.30	543.01	543.00	0.0007	22	
CR3N3.1	CR3N4	39	400	0.30	542.99	542.99	0	0	
CR3N4.1	CR3N5	10	390	0.30	542.95	542.83	0.012	92	
CR3N5.1	CR3N7	50	340	0.30	542.83	542.81	0.0004	17	
CR3N7.1	CR3N8	48	292	0.30	542.83	542.88	–0.001	0	Backfall
CR3N8.1	CR3N9	36	256	0.30	542.89	542.38	0.0142	100	
CR3N9.2	CR3N10	13	243	0.30	542.38	542.38	0	0	
CR3N10.1	CR3N11	42	201	0.30	542.34	542.35	0	0	
CR3N11.1	CR3N12	1	200	0.30	541.57	541.62	–0.05	0	Backfall
CR3N12.1	CR3N13	1	199	0.30	541.61	541.59	0.02	119	
CR3N13.1	CR3N14	20	179	0.30	541.58	541.60	–0.001	0	Backfall
CR3N14.1	CR3N16	28	151	0.30	541.56	541.57	0	0	
CR3N16.1	CR3N17	19	132	0.30	541.56	541.61	–0.002	0	Backfall
CR3N17.1	CR3N18	28	104	0.30	541.58	541.52	0.0021	38	
CR3N18.2	CR3N20	104	0	0.30	541.43	541.38	0.0005	19	

Notes: [1] Pipe numbering convention is to use upstream MH ID followed by '.1' for first outlet pipe, '.2' for second outlet pipe, etc. In most cases, there is only a single outlet pipe. This convention avoids the need to list the upstream node in table.

[2] Capacity is based on Manning Equation (Chapter 8). $Q_{l/sec} = 1000 \times Q_{m3/sec} = A \times V = A_{m2} \times 1/n\ (R^{2/3})\ s^{0.5}$

For full flow in pipe of diameter D, R= D/4. The equation becomes $Q_{l/sec} = (312/n) \times D^{8/3} \times s^{1/2}$ (D in metres, and s in m/m).

Table 9.4 Detailed road and kerb survey sheet

Site: Motilal ki Chal
Date: 8/8/94
Surveyor: Dilip Sharma

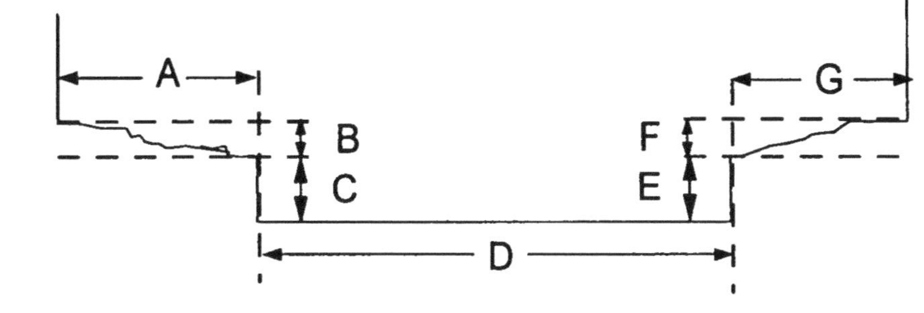

Street	From	To	A North or West Margin Width (m)	B North or West Margin Height (cm)	C North or West Kerb Height (cm)	D Road Width (m)	E East or South Kerb Height (cm)	F East or South Margin Height (cm)	G East or South Margin Width (m)	Remarks
R1	Main	CR1	3.45	0	7	4.10	8	0	4.90	
	Road		5.00	10	7	4.90	7	15	5.00	
			5.10	15	7	3.95	8	15	5.15	
	CR1	CR2	4.90	15	7	4.10	7	15	4.85	
			4.70	10	7	4.10	7	10	5.25	
			5.00	15	7	4.10	8	15	4.70	

Note: These data were collected to assist in approximate assessment of the capacity of the road network as a drainage system. While this chart can give some rough idea of capacities, detailed profiles of the road surface would be required to allow for the effect of camber in reducing capacity.

number of any manhole; this is an invitation to disaster in later review of raw data, as it becomes unclear which 'version' of the number was meant. Thus, gaps appear in some of the tables, as 'incorrect' manholes originally considered were removed.

Condition of conduits

Open conduits

It is relatively easy to inspect open drains to determine their structural condition, slopes, and major blockages. The easiest but rarest case is that of a clean surface water drain containing no sewage or sullage in which simple visual inspection of the channel is sufficient. Where solids cover the bottom of the drain, these must be cleared at intervals to measure the depth and assess the structural condition of the bottom. The survey should note such details as obvious collapses or bulging of channel lining, loss of mortar or bricks, major cracking, etc. Where the drain carries sewage or sullage, inspection is more difficult, and is best done during the lowest daylight flow.

Open drain systems are rarely open along their entire length; there are almost always sections which are covered by street crossings or encroachment. These locations are worth noting, particularly where they restrict access for cleaning. Measuring the length of channel to which a cleaner does not have easy direct access (e.g. covered sections longer than three metres) can sometimes reveal that apparently 'open' channel systems are in fact effectively closed along a third of their length. Such systems have the worst of both worlds: on the one hand solids from the surface can easily enter them, but on the other maintenance is complicated by the covered parts which are hard to clean.

Closed conduits

It is much more difficult to evaluate the condition of closed conduit storm drains. As for open drains, inspection is easiest where drains are truly separate. Even in this case, however, unless the closed drain is very large, or manhole spacing is unusually frequent, it is difficult to assess the condition, except by spot checks at manholes. In the industrialized world, closed circuit television (CCTV) technology for drainage assessment has come into its own over the past two decades. Where it can be afforded and managed, it represents the best technical solution to the problem of assessing the condition between manholes. Unfortunately, such an option is beyond the reach of most municipalities in developing countries. Some methods are suggested below as a start on this difficult problem; even where CCTV is available, some of these techniques may be useful in setting priorities as to where it can best be used.

Manholes are the obvious place to start to assess the condition of pipes. Manholes are almost always cramped, uncomfortable, dirty, and wet; they are sometimes dangerously deep. Methods which minimize the inconvenience and hazard of manhole entry are therefore preferred. In the end, however, there is frequently no other way to check an invert level or assess what is going on than simply to get in the manhole and have a look.

Safety is always an issue when working in drains, and especially when entering deep manholes. Certain basic precautions must always be observed in this work. These include:

- All workers in drains should be wearing basic protective clothing, including gloves and boots to reduce risks of contamination. Thick gloves are particularly important when handling solids from drains, as these often contain glass.
- Any cuts should be cleaned and treated immediately to prevent infection.
- Nobody should work in a closed drain alone. Co-workers are important in case there is a need for help in an emergency. The co-workers may be on the surface, but should keep the worker in the drain in sight or earshot at all times.
- Ensure adequate ventilation in closed drains. When entry into a deep combined sewer is required, open up the entry and adjacent manholes and allow half an hour for sewer ventilation before entry.
- Nobody should enter a deep manhole without a harness. Crews should practise lifting workers from the surface using the harness, so that they know what to expect and how to perform in an emergency.
- Keep tools and equipment clear of the manhole entry. Spanners and crowbars give little warning as they fall, and the workers in the manhole have very little room for manoeuvre.
- Nobody should start work in a deep manhole during rainfall. Rainfall can shift from light to intense with little warning, and once flows are developed, water levels can rise suddenly. When rain begins, the priority should be to leave the work in good order and get out.

Extensive safety regulations for work in enclosed spaces are enforced in many industrialized countries, but these may require equipment (e.g. breathing apparatus) that is not readily available in developing countries. The simple principles outlined above, however, require only common sense and care.

Standing water checks In a separate storm drain, one very quick check for possible blockages is to look for standing water above the invert of the outlet. Such standing water is a clear sign of a blockage, collapse, or a backfall on the pipe; very simply, it shows that the pipe does not drain properly (Figure 9.3).

Lamp-mirror survey If a pipe is clear and straight, a powerful lamp lowered in one manhole should be visible at an adjacent manhole with a mirror as shown in Figure 9.4. This technique not only has the advantage of reducing entry from above, but the mirror often provides better visibility in small manholes where there is little or no space for the field worker below the pipe

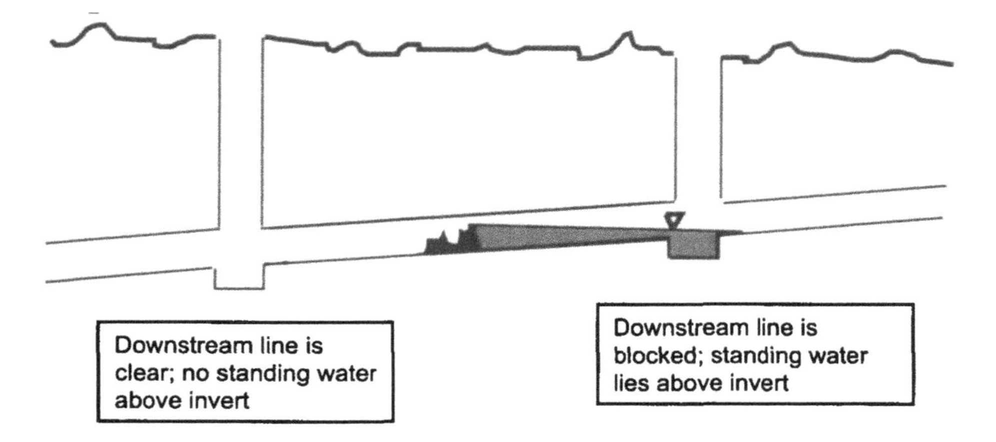

Figure 9.3 *Standing water check as a sign of pipe obstruction*

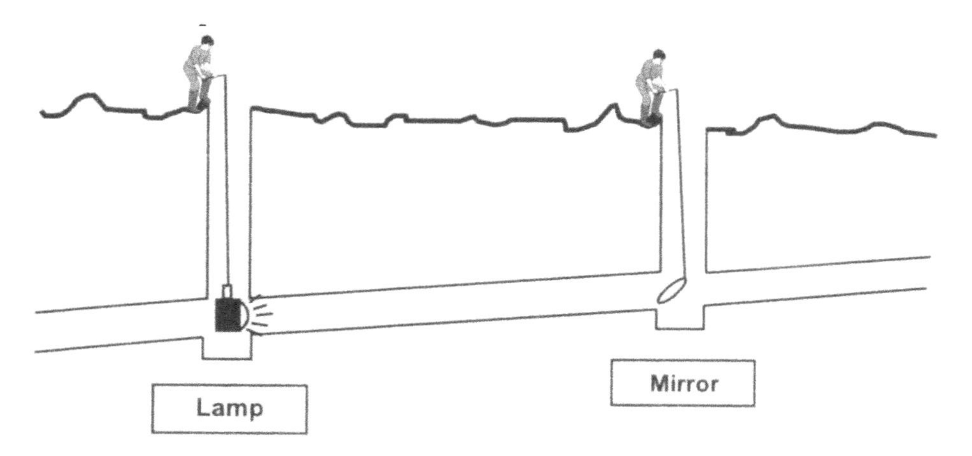

Figure 9.4 *Lamp-mirror survey*

outlet. The lamp must be a powerful one, and care must be taken to protect the batteries and wiring from moisture. A form for this type of survey is shown in Table 9.5.

Cairncross and Ouano (1991) describe how such a survey could be used to show a variety of problems: for example, a sagging pipe will obscure the top of the lamp's image in the mirror. In practice in Indore, we found such diagnosis very dependent upon the length of the run and the size of the pipe. The method was particularly helpful in identifying problems on short runs of pipe where blockages became obvious. On longer lengths, it is more difficult to interpret the extent or nature of a problem. In summary, lamps and mirrors can help and are a good start on the problem, but much of the time they are not enough to explain the cause or type of blockage.

Table 9.5 Lamp-mirror survey form

Date: Aug. 3, '94
Surveyors: Tiwari & Kolsky
CR3 northern line survey

U/S Manhole	*D/S Manhole*	*Length Centre to Centre*	*Light Visible?*	*Clear or Broken?*	*Depth of water in pipe?*	*Water standing or flowing?*	*Visibly clear area of pipe?*	*Remarks*
CR3N3	CR3N4	35.8	Yes	Clear	<1/4	flowing	>90%	
CR3N7	CR3N8	46.8	Yes	Clear	1/4	flowing slowly	3/4	
CR3N10	CR3N11	44.9	No	–	<1/4	no flow	–	
CR3N13	CR3N14	13.4	Yes	Clear	<1/4	No flow	>90%	Pipe is clear above water
CR3N16	CR3N17	16.2	Yes	Clear	<1/4	Standing		Lamp not bright enough to see much; batteries fading
CR3N20	OUTLET	33.3	No	–	1/2 u/s <1/4 d/s	Standing u/s Flowing d/s	0	Visible rubbish at u/s end
CR2R7	CR3R7	55.1	Yes	Barely Visible			All visible at outlet	Distance too great for this lamp

Manhole inspections Often there is no alternative but simply to get down the manhole and have a look, particularly where a problem is suspected; this is why manholes were built! Most drains are too small and low to allow field workers to look along them from the manhole, so it is helpful to carry a small mirror to reflect the view from the mouth of the pipe. As above, a lamp lowered down the neighbouring manhole is useful to illuminate the pipe from the other end and identify blockages. A good small camera is also an invaluable tool; it can take a picture (with flash!) of pipe sections that are physically impossible for a field worker to observe directly. Each picture needs to be carefully documented in a notebook, indicating the line being photographed and from which manhole. It is worth taking the time to sketch the situation, as otherwise there will be confusion about which pipe has been photographed from where. Cameras which can include date and time information on prints are handy for *all* aspects of drainage evaluation. Unfortunately, modern battery-powered cameras are particularly susceptible to moisture; older models may be more practical, but a flash attachment is essential.

Apart from the pipes themselves, it is worth looking at the condition of the manhole, and any solids that may be in it. If the manhole itself is in bad structural condition, this can lead to blockage of the pipe either from the debris of the crumbling manhole, or from the entry of the soil which the walls should be holding back. In some cases, it will be obvious that the manhole is being used as a dump for solid waste; in other cases, particularly in the first rainy season after installation, shuttering and formwork may be left inside from construction.

Manholes are frequently built where other pipes or sewers cross the drain. These can represent major obstructions to flow, which will rarely be apparent from design drawings, and can only be determined from a manhole by manhole survey of the network. The problem can be particularly acute where separate storm drains and sanitary sewers have conflicting demands for the same underground space. Quantifying the effect of such blockages on hydraulic capacity is difficult; this is, however, no reason to ignore them!

Manholes are hydraulically distinct from pipes; where benching is not provided to guide it, the flow expands and slows down after leaving the upstream pipe and entering the manhole. This means that solids often settle out in manholes that would otherwise carry on through the pipe. It is worth looking at this debris, to see if there are any clues to upstream damage. Where clay pipes are used, for example, pieces of tile found in a manhole are often a sign of upstream problems from tree roots or other damage to pipes. Alternatively, the debris may indicate the break-up and entry of road surfacing materials, or the entry of soil through broken pipes or unsound manholes. It may not be clear *where* such debris has entered the system, but it can point out problems to watch for.

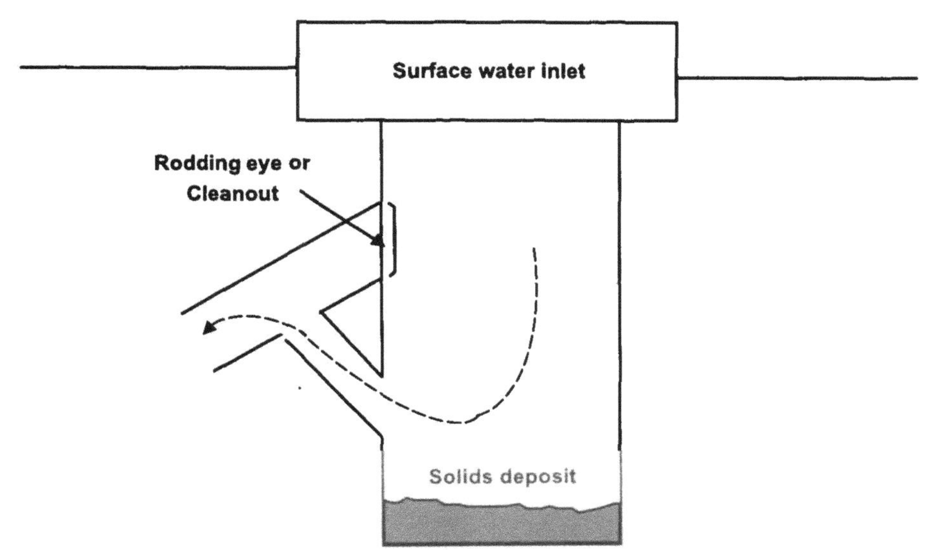

Figure 9.5 *A gully-pot installation to trap solids before their entry to the drain*

Condition of inlets

Different networks use different types of inlets, and it is worth documenting what kind are in use, and any problems they cause. Here, conversations with maintenance staff are critical; they will know the difficulties of maintenance only too well! Some systems have inlets with gully pots to trap solids before they enter the drain, as shown in Figure 9.5. Such inlet structures may be connected directly to the main storm drain, or indirectly via a manhole. Still other systems use direct connections to the pipe or manhole without traps.

Relevant questions when inspecting inlets and interviewing maintenance staff include:

- Are inlets damaged or broken?
- Are they vulnerable to damage from traffic?
- Are they well-located?
- Are they set at a suitable level to drain a significant area?

Many of the ideas mentioned above for manhole inspection apply to inlets; it is worth removing the debris from an inlet, drying it out, and observing the sorts of things that enter the inlet. Similarly, lamp and mirror tests may reveal problems or blockages between inlets and the downstream manhole if the inlet is not connected directly to the main storm drain.

10 Maintenance surveys

DRAIN MAINTENANCE CONSISTS of (1) the periodic removal of solids deposits and blockages from the network, and (2) the performance of routine repairs. The previous chapter has outlined some clues in assessing structural condition; local engineering and practical building experience will be a sound guide as to how well routine structural maintenance has been carried out. This chapter describes approaches to the monitoring and evaluation of solids deposits in drains.

Drain solids surveys

Solids are much easier to assess in open drainage systems than in closed conduit networks. Open drains offer better access to see what is going on, to measure levels, and to sample the solids. In closed conduit systems, inspections of solids at manholes may give some qualitative insight into the kinds of problems the system may experience; solids levels in manholes, however, are not representative of levels in the pipes themselves. The solids of interest are those in the actual conduits, and there is no easy way to assess their 'depth' in pipes too small to enter. Most of the methods described below, therefore, apply only to open channel systems.

Visual observation

The first step is to walk along the network to identify obvious blockages, or points where solids levels seem relatively high. These may be apparent as in Figures 10.1, 10.2, and 10.3. A quick inspection after a heavy storm can be particularly revealing, as small projections (e.g. a small water pipe across the drain), which might pass unnoticed in dry weather can trap large amounts of debris and become a real obstacle to flow. (See Figures 10.4a and b.)

Solids levels

In open channel systems, solids levels can be measured quite easily. If the measurement is going to be done only once, then the simplest way is to clean out the drain at the point where measurements are needed, and measure the depth of the adjacent solids deposits. Where this varies by more than 30 per cent across the section, it is worth recording values taken at three points, e.g. at the centreline, at a point one-quarter of the way across the channel's width, and at another point three-quarters of the way across the channel.

http://dx.doi.org/10.3362.9781780446059.008

Figure 10.1 *Mud completely blocks this drain that was originally over 30 cm deep*

Solids build-up surveys

Apart from the level of solids in the drainage network at a particular time, it is worth knowing how *quickly* the solids accumulate. If a clean drain gradually fills up over a year, then an annual cleaning, just before the onset of rains, is adequate. If the drain fills up in a month, then the drains need to be cleaned more frequently. If slopes are good and solids don't accumulate, then cleaning need not be scheduled at all, and can be done only in response to complaints as and when they arise.

To measure the rate of solids build-up, weekly measurements of solids levels are required. These can easily be done using the apparatus shown in Figures 10.5 and 10.6. (The crossbar can be marked symmetrically from the centre to ensure that measurements are made at the centreline and thus at the 'average' height of the channel sides.) This equipment measures the depth

Figure 10.2 *Residents have blocked these inlets to protect cats, dogs, and small children; engineers need to respond to such concerns*

Figure 10.3 *Rubbish and construction debris have accumulated at this point to more than half its depth*

Figure 10.4 (a and b) *This small pipe seems innocuous in the dry . . . but catches debris and obstructs flow in the wet.*

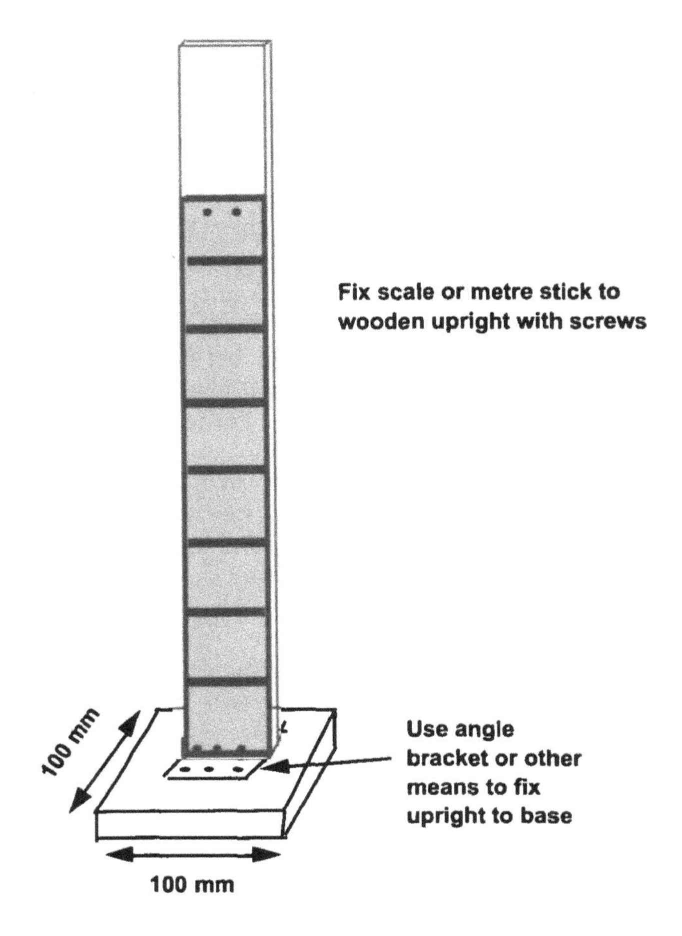

Figure 10.5 *Scale for solids measurement*

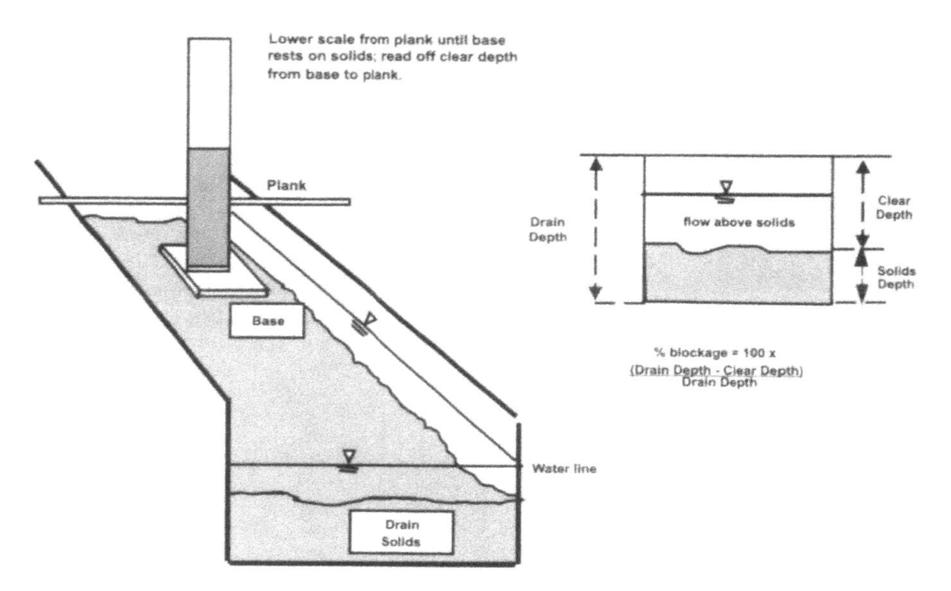

Figure 10.6 *Use of the solids scale within a drain*

from the top of the channel down to the level of the solids. To compute the depth of the solids, the total depth of the channel must be known. As with the single measurement of solids depth, this will require complete cleaning of the drain at the points where such measurements are made. While it may seem easy to push a rod through the sediment to the bottom of the drain, we found that much of the time the rod will stop at something solid like a rock or brick; the only way to know the depth of the channel bottom is to clean out the drain at this section.

These data can be recorded on the form shown in Table 10.1. Solids depths, and the fraction of area blocked by solids can be rapidly computed along the drain with a computer spreadsheet package. The solids depth is equal to the total depth minus the clear depth measured with the scale.

Solids sampling and size distribution

In addition to knowing the levels of solids in the drains, it is helpful to know the size distribution and approximate composition. If, for example, the solids are very fine, then they are more likely to be swept out in the first flush of the next storm. If they are present after a storm the night before, then they show a low velocity of flow. This may be because the design slope of the drain is too flat, or it may be that the downstream water level 'backed up', and restricted the upstream flow. Large silt deposits are often found in sanitary sewers when the operation of downstream pump stations is limited (e.g. at night to reduce power costs). This stoppage of flow reduces the velocity to zero, allowing the settling out of fine particles that can then consolidate into a sticky immovable

Table 10.1 Data form for solids surveys in open drain

Sheet No 1 of 2

Site: Pardeshipura
Surveyor: M. Ali
Date: 27/9/94
Weather:
  last night: Dry
  this morning: Dry

General Remarks:

Name of drain: Main Drain

Station No.	*Time*	*Clear Depth (scale reading)*	*Flow*	*Type of solids & floating matter*	*Remarks*
0	11:57	94	Med	One boulder on side	
1		88.5	Med	Clear	
2		90.5	Med	Sand	
3		96	Med	Rubbish, stones	
4		62.5	Slow	Rubbish, mud, sand & stones	
5		65.5	Slow	Mud, sand & stones	
6		54	Slow	Mud, sand & stones	
7		64	Slow	Mud & sand	
8		59.5	Slow	Mud, sand & stones	
9		51	Slow	Mud, sand & stones	
10	12:08	53.5	Slow	Mud & sand	
11	12:10	49	Slow	Mud, sand & stones	
12		56	Slow	Mud, sand & stones	
13		52	Slow	Mud, sand & stones	
14		51	Slow	Mud, sand & stones	
15		51	Slow	Mud, sand & stones	
17		29	Slow	Rubbish, mud, sand & stones	
18		42	Slow	Mud, sand & stones	
19		55	Slow	Mud, sand & stones	
20		38.5	Slow	Rubbish, mud, sand & stones	
21		56	Slow	Rubbish, mud, sand & stones	
22		41.5	Slow	Mud, sand & stones	
23		40.5	Slow	Rubbish, mud, sand & stones	

mass. Large silt deposits show inadequate hydraulic slope, which can be managed only by either reducing the downstream water levels, or relaying the drain at a steeper slope. If the downstream water levels are too high, increasing the slope of the conduit will make no difference; it is the *difference in water levels* which controls the velocity of flow, not simply the slope of the conduit.

Alternatively, when solids are very coarse, lowering the downstream water level or increasing the slope is unlikely to help very much. In this case, the only solutions are:

- to keep the solids from entering the system
- to trap the solids after they enter the inlet, but before they enter the conduit (e.g. through gully traps)
- to clear the solids from the drain, or
- to develop alternative drainage routes (e.g. surface routes).

To sample for sediment size, the best approach is to pick a uniform spacing along the drain. Avoid sampling at unusual locations (e.g. culvert inlets and outlets), unless you are particularly interested in what is going on there; these points will probably be different from most of the drain length. At each of the sampling points, clean out the drain entirely for some specified length (e.g. 1 metre) and deposit the cleanings into buckets with covers to transport to a soils lab. The soils lab can then dry the samples and do a particle size analysis. A simple sieve analysis should be adequate to give an idea of the physical transport issues in the system; this cannot tell you about the distribution of fine sediments, but where most of the solids are above 1 mm in diameter, there is little point. In the open channel catchment we studied most closely in Indore, it was apparent that the main problems were with the large solids, not the small ones. The results of such a size analysis were shown in Figure 2.6, and compared with those obtained from a site in Britain. These Indore results were averaged for five sampling points; it may be useful to compare results at different locations to see if there are characteristic differences.

Inlet solids surveys

Two different sets of observations are worth making on the blockage of inlets.

Blockage of inlet mouths

Blockage of inlet mouths can be seen from the street quite easily. There is little point in precise measurement, but there is a lot to be gained by estimating to the nearest quarter; if over 50 per cent of the inlets are more than half blocked, for example, inlet maintenance is much more of a problem than if only 25 per cent are more than one-quarter blocked. Using such 'quarter measurements', you can plot up charts such as Figure 10.7. Note that in this case, a year after construction, about one in three of the inlets is at least half blocked, and one in six is at least three-quarters blocked!

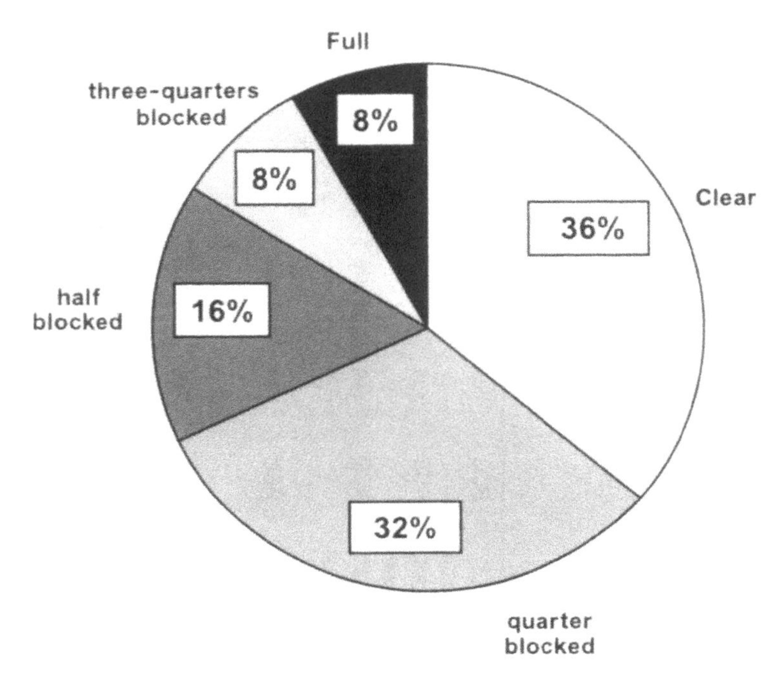

Figure 10.7 *Distribution of blockage of street level inlets; survey of 25 Inlets, 30/6/94, Bhagirathpura, Indore*

Solids levels

Measurement of the level of debris inside the inlet can show problems not evident from simply looking at the mouth. Where there is a gully pot acting as a trap, it is useful to know if the solids either block, or are close to blocking, the connection to the drain itself. Where the inlet consists of a direct connection to a manhole, then it is worth recording the level of solids relative to the inlet and outlet pipes. Levels are recorded from the crown of the pipe, as they cannot be measured directly from the invert when it is covered. Table 10.2 shows one way of recording this used in Indore. Note that this form shows data for both the solids levels in the manhole and for the blockage of the mouth of the inlet. These surveys can be done separately; in particular, surveys of the inlet mouths do not involve the time-consuming requirement to open the manhole for making these measurements.

Drain cleaning observation

Drainage maintenance staff know a great deal about the maintenance problems of the drains on which they work. They should be asked about the problems they face on the job, any deficiencies in tools and equipment, and

Table 10.2 Solids levels in storm inlet manholes

Date: 30/6/94 **Location:** Bhagirathpura
Time: 11:10 AM **Surveyors:** M. Ali, M. Haidry, S.Tiwari

Drain & Inlet ID	*Main Inlet Pipe Size (mm)*	*Outlet Pipe Size (mm)*	*Main Inlet Crown to Solids (cm)*	*Outlet Crown to Solids (cm)*	*Mainline Inlet: fraction blocked (¼, ½, etc.)*	*Tributary Inlet: fraction blocked (¼, ½, etc.)*	*Outlet fraction blocked (¼, ½, etc.)*	*Type of Solids*	*Type of cover*	*Cover In place?*	*Fraction of blockage of street inlet (¼,½, etc.)*	*Remarks*
S-8/1	300	300	30	20	Clear	–	<¼	silt, sand & stones	RCC	Yes	Clear	
S-8/3	300	300	18	18	¼	Clear	½	Rubbish, silt, sand & stones	RCC	Yes	¼	
S-8/4	300	300	38	30	Clear	Clear	Clear	Rubbish, silt, sand & stones	RCC	Yes	Fully	Inlet blocked by boulders placed by residents
S-8/5	–	300	–	16	–	–	½	Rubbish, silt, sand & stones	RCC	Yes	¼	
S-8/6	300	450	31	45	Clear	Clear	Clear	silt, sand & stones	RCC	Yes	Clear	
S-8/7	None	300	–	0	–	–	Fully	silt & sand	RCC	Yes	½	inlet manhole below unstable soil pile
S-8/8	450	450	30	30	¼	Clear	½	Rubbish, silt, sand & stones	RCC	Yes	Clear	
S-8/9	300	300	?	?	Fully	Fully	Fully	Silt	RCC	Yes	¼	Solids level above pipes in MH

those aspects of design, construction and maintenance which cause them the greatest difficulties. There are two particular aspects which should be directly observed in any systematic evaluation.

Removing solids from the drain

Without making too much fuss, it is helpful to visit a crew doing a routine maintenance job and see how they do it. What tools do they use? What protective clothing do they wear (e.g. gloves and boots)? What precautions do they follow? Are there convenient washing facilities for the workforce? There is almost always some contamination of surface drains with human or animal wastes, and the solids deposits at the bottom of combined drains are heavily contaminated with human wastes. In addition, there are often 'sharps' (e.g. glass and razor blades) which can easily cut through thin gloves and create the risk of infection.

From the drain to safe disposal

In many places, removing solids from the drain is less than half the story. Drain cleanings can often sit beside the drain for months. An initial delay of one or two days may be reasonable to allow some drying before transport, but this period should be kept to a minimum. In practice, the delay often becomes indefinite as the transport required to remove the solids for safe disposal is unavailable or not well organized. In many cases, these drain cleanings contain high concentrations of parasitic eggs (e.g. roundworm or hookworm eggs) that can easily infect children playing in the area. As days go by the piles of cleanings become less distinguishable from the surrounding soil. There are therefore several questions to ask about the later stages of drain cleaning:

- How long does it take between the cleaning of the drain and the removal of the cleanings from the side of the drain?
- What precautions are taken to protect the community (especially children) from these concentrated wastes?
- Where are these cleanings taken and does this present a hazard?
- What precautions are taken by the work force in picking up and transporting these wastes?

If the evaluation shows that solids in drains represent a serious reduction in drainage capacity, it is well worth thinking about these questions before initiating a major drain cleaning effort. These points should be discussed with the maintenance crews and managers in a practical constructive way. There are often no easy answers to these questions, and a realistic objective of most evaluations must be to improve things incrementally, rather than to solve problems completely.

Apart from the actual cleaning procedures, it is straightforward to monitor the piles of drain cleanings themselves, simply by marking their location on maps, and passing by every other day or so to see if they are picked up. The forms for doing this are similar to those for solid waste monitoring, described below.

Solid waste monitoring

Where solid waste seems to be a likely source of drain solids, it is helpful to learn about the solid waste management system. In many cases, solid waste may be brought to collection points; if these are near an open drain or an inlet to the drainage network, solid waste can easily enter the drain. If solid waste collection is poor or irregular, residents will be forced to find other ways to dispose of their rubbish; in such cases, it is likely that it will end up in the drain. Piles of solid waste can be mapped and monitored, just as we have suggested for drain cleanings above, although piles of solid waste are usually picked up more frequently. As with flood monitoring, community members can be helpful, and can explain some of the problems that arise from irregular collection. As with the drain cleaning work, it is also useful to talk with the solid waste workers to understand the difficulties under which they work. Table 10.3 shows a form used for monitoring rubbish pick-up in Indore.

Table 10.3 Extract from solid waste monitoring

Date: 23/10/93 **Surveyor:** S. Tiwari
Time: 11 AM – 1 PM

Pickup Point Number	*Approx. size of pile (Diam × height or L × W × h)*	*Remarks*	*Changes from previous day*
W-8/1	1.5 Diam X 0.4 m	Rubbish covered with a soil pile.	No change
W-8/2	2.5 Diam × 0.3 m		Size increased
W-8/3	3 m × 2 m × 0.2 m	Scattered rubbish	No change
W-8/4	3 m × 2 m × 0.2 m	Scattered	Size increased
W-11/3	2.5 m Diam × 0.3 m		Size increased
W-11/2	3 m × 3 m × 0.1 m		Size increased
W-13/1	2.5 m × 2 m × 0.2 m	Scattered	Very small increase in size
S-13/1	No pile		No accumulation, since start of festival
W-19/1	2.0 m × 0.3 m		Picked up yesterday by IMC workers.

In addition to monitoring the piles, a 'surface solids survey' can show the kinds of solids that are likely to enter the drain. Such a survey is simply a walk along the drainage channels in the catchment, identifying possible sources of solids that can contribute to inlet and drain blockage. These can include piles of rubbish, stockpiles of construction materials next to the drain, or mounds of unstable soil adjacent to a drain inlet. Such a survey will not alter estimates of capacity, but can help to explain what kinds of materials are entering the drain, and how difficult it may be to prevent it. This survey is not worth repeating regularly, but it is instructive to perform once early in the study. Table 10.4 shows one of these forms used in Indore.

Table 10.4 Extract from street solids survey

Date: 21/8/93 **Surveyor:** Tiwari
Time: 3–5:30
Site: Motilal ki Chal

Street	*From*	*To*	*Possible blocking materials*	*Location*
R3	CR4	CR3	A brick pile 2 m × 2 m × 2 m	on East Side, crossing of CR3 and R3
			Aggregates & brickbats, pile 1 m Diam × 0.5 m height	near large brickpile above
	CR3	CR2	Small brick pile 1 m × 1 m × 0.5 m	West side, 15 m from CR3
CR4	R2	R3	Pile of rubbish & loose soil, 3 m × 1 m × 0.3 m	On north side, 10 m from R3
			Gravel, rubbish scattered in area 3 m × 1.5 m	
			Scattered murram, 6 m × 3m × 0.5 m	S. side, 0.4 m from R3

11 Studying drainage systems in action

THIS MANUAL HAS stressed the need for engineers to think about how catchments and drainage systems *actually* behave during storms. The best way to start is to observe the catchment and drainage system during storms. This is difficult for a variety of reasons, and generally does not yield the same kind of numerical data as the surveys of the previous two chapters. Wet weather work, however, promotes qualitative technical insight that is simply not obtainable any other way. Specifically, wet weather studies are the most effective way for the engineer to assess:

- catchment and subcatchment boundaries
- the nature of flooding in flood-prone areas
- the hydraulic performance of the total drainage system
- the surface flow routes followed by runoff during floods
- the public nuisance, hazard, and damage associated with flooding.

Technical staff can also measure the quantitative variations in water level with time at a number of locations during a storm. In detailed hydraulic modelling studies, such measurements are an invaluable complement to electronically recorded data, as they permit collection of more data at relatively low cost. The rapid variations in flow and water level during a storm, however, make such measurements virtually impossible to interpret without the use of one of the powerful hydraulic models described in Annexe 8-A. Unless a full-blown dynamic hydraulic analysis is intended, technical staff are more usefully employed assessing the qualitative performance issues described in this chapter.

There is tremendous difficulty as well as great benefit in doing effective wet weather observation. The work requires clarity of purpose, strong organization, and strong team motivation and discipline. These requirements stem from:

- the unpredictable onset, duration, and nature of storms and
- the chaotic and unpleasant nature of working in floods and storms.

Very simply, nobody knows when a storm will occur, how long it will last, or how intense it will be. Under these conditions, it is essential that staff members at least know what they will do and where they need to be at the onset of rain. This chapter offers some suggestions on how best to organize wet weather observations.

http://dx.doi.org/10.3362/9781780446059.009

What to look for in wet weather

Catchment and subcatchment boundaries

In areas with reasonable slopes, good mapping is adequate to define the boundaries of the overall catchment. Many flood-prone urban areas, however, are relatively flat, and the surest way to define tributary areas and the directions of flow is by observation. In particular, there are two distinct types of observations that need to be made:

- *During moderate storms*, a simple survey using the catchment maps described in Chapter 5, on which the direction of flow (or ponding areas) can be marked. In flat areas, minor variations in levels can lead to substantial changes in tributary areas, and flow may enter the drainage network at unlikely locations. For closed pipe systems, for example, some inlets may end up intercepting little or no flow because of irregularities in topography.
- *During events with significant flooding*, note how the overall flow patterns and tributary areas have altered as a result of flooding. In many catchments, the drain itself defines a boundary of the catchment. In this case, when the drain floods, the overflowing water may leave the catchment, and enter an adjacent one. By the same token, other catchments – which are hydraulically separate under ordinary circumstances – may overflow into the one under study during a heavy flood. These effects are worth noting, both for evaluating the hydraulic capacity and requirements of your network, and for understanding the consequences of flooding.

The nature of flooding in flood-prone areas

The resident surveys and discussions described in Chapter 6 should offer some idea of where to look to observe the extent of flooding. Such surveys, however, may not give a good idea of the extent of any particular problem area, the bottlenecks which cause the flooding, or the difficulties in drainage of the flood waters. Observation is also the best way to get a sense of whether some of these areas flood a great deal more frequently than others.

Unless these areas have been identified beforehand, however, systematic observation is difficult. While a roving survey of the entire catchment during a storm can be useful to assess performance, it is not effective in identifying flood-prone areas, as the flooding will vary in both time and place; observers will inevitably miss some of the action while moving from one area to another. If a study of flood-prone areas is desired (e.g. to consider options to minimize the problem), it is essential to define these areas geographically and assign specific team members to observe them during storms. Observation of flood-prone areas often clarifies the cause of the flooding, such as inflows from other areas, or inadequate inlet capacities.

The hydraulic performance of the total drainage system

The clearest way to see the differences between the simplified hydraulic theory of capacity analysis and the hydraulic performance of drainage systems is a systematic survey of the drain itself during storms. Such a survey can establish:

- overflow locations
- bottlenecks and high head losses, e.g. culverts
- obstructed entry to the drain, either by inlet blockage, poor inlet design, or poor surface grading.

Note that the drainage 'network' is more than just the drain; it includes surface and gutter flow as well as inlet and drain flow.

The surface flow routes followed by runoff during floods

When flooding takes place, flood waters usually follow routes other than the drainage channel to escape. These need to be noted both to consider the impacts on affected residents, and to consider options to mitigate problems. In some cases, flow simply leaves one drain, travels down a street and re-enters another with virtually no effect on residents. In other cases, whole areas become inundated. Understanding what happens in these cases helps to set priorities on what needs to be done.

The nuisance, hazard, and damage of flooding

Residents know better than anyone how much of a nuisance flooding is; observation during storms is no substitute for talking with people. Chats with residents just after floods will suggest some of the problems to watch for; in fact, residents may well seek out team members *during* floods to demonstrate the problems. While at the time this may seem a distraction from the other work to be done, it is at least as important.

How to manage wet weather tasks

Organizing a team

Storms are unscheduled, chaotic, and physically unpleasant opportunities for learning. To make the most of wet weather, observation must be organized and systematic, and staff must be well disciplined to carry on tedious work in difficult conditions. Effective wet weather work, and analysis of its results, also requires an understanding of the catchment and the drainage system. For these reasons, it makes sense to set up fixed teams for wet weather observation, rather than hoping that different people can perform equivalent observations in different storms. Staff who do the same job in different storms are quick to pick up differences in drainage performance between the events.

To be effective, this group must develop a sense of pride in the work. This may be encouraged in a small way through the project's acquisition of good (and distinctive!) wet-weather clothing and waterproof backpacks along with basic equipment such as measuring tapes and clipboards. A case can be made for offering a small bonus payment for each storm in which members participate, subject to the satisfaction of the manager of the evaluation. Such a bonus is awarded individually, and for performance; it is not a collective 'right' of team members.

It is worth setting aside half a day after each storm to allow for debriefing and the transfer of notes from rough field notes to final copy. Wet weather observation helps much more with qualitative insight than quantitative data collection, but this requires sharing the experience of team members working in different parts of the catchment or on different tasks. It is also worth gathering team members together to discuss working methods: forms can often be simplified after team members point out the difficulties of filling in too much data in the middle of a thunderstorm! The original rough sheets must, however, be kept, as every time data are copied, the risk of transcription error increases.

The quality of field data depends largely upon preparations made before the onset of the rain. Some way of writing in the rain must be developed; we found soft pencils worked well on plastic-impregnated fabrics, which were readily available in the market for other purposes. As part of the debriefing after each storm, the tasks and responsibilities for the 'next storm' should be determined and allocated by the manager. If this is done properly, team members know where they have to go and what they have to do at the next onset of rain *without* having to assemble as a group.

Team members can usually take on more than one task per storm. Work is most effective when team members have a clear idea of priorities and sequencing of tasks. A typical set of instructions can be something like: 'Start with tracing the direction of flow in the area outlined on your map; when you've finished that, move on to checking the inlets in your stretch of the drain; after that, check to see what's going on down at the corner of Street X and Y, and if there's flooding, see where the water goes.'

While collaboration is essential in debriefing, team members in the field should work as independently as possible. There is a strong temptation to 'double up' in inefficient teams of two or three, where twice as many sites or jobs can be performed if staff members can work as individuals. Where difficult measurements need to be made and recorded, there may be a good case for working in groups of two, but even here they do not need to be of the same professional level; in many cases a technician and a labourer can get more done than two technicians.

Organizing specific tasks

Checking catchment and subcatchment boundaries This job is made much easier if good maps have been prepared as part of the work described in Chapter 5. Team members should each be allocated a 'beat' within the catchment, and simply note on suitable copies of their maps the direction of flow during the early portions of the storm. This should be completed within the first storm or two of the season, as it is essential to understanding the limits of the catchment. This activity may identify surprising differences with preconceived notions about the catchment which can be usefully followed up later; for example unsuspected cross-connections between sanitary and storm sewers may alter catchment definitions substantially. In addition to the determination of flow routes *within* the catchment, an effort should be made during a very heavy storm to see if the overall catchment boundaries hold, and to what extent runoff crosses catchment boundaries during a flood.

The nature of flooding in flood-prone areas This work is similar to that of determining catchment boundaries and the direction of flow; assigning specific beats or areas to individuals is more useful than any individual trying to observe flooding in more than one place. One difficulty is that just as storms are unpredictable, flooding itself is even more unpredictable, and being in the right place at the right time is difficult. It is for these reasons that the chalk gauge and resident gauge methods were developed (see Chapter 6) and are more likely to give a reasonable idea of the extent of flooding.

Forms, however, can be developed to identify the nature of flooding in these areas when it occurs. Field workers can examine, for example, the rough limits of the ponding area; what fraction of houses have plinth levels covered in water, and whether slow drainage of flood waters is attributable to poor inlet capacity or location. The difficulty is in being at the right place at the right time; such logs are likely to place lower limits on the problem of flooding.

Hydraulic performance of the total drainage system Hydraulic performance can be limited by the drains themselves, poor entry conditions into the drain, and restricted surface flow towards the drain. Structured visits along the length of the main drainage network during storms are useful for identifying hydraulic problems in the drains themselves, and in the entry of runoff into the drain. Where the drainage system is extensive, it can be useful to divide the drainage network itself into sections which can be managed by individual field workers. The team leader, however, should try to do one complete traverse of the main system during at least one storm, and preferably more.

Observations of hydraulic performance of drains and inlets are best done when the system is not flooded. In practical terms of sequencing work, this means studying drains and inlets at the onset of rain or after the peak of a flood; different types of blockage can become apparent during these two stages, as flood waters may make the problem worse by transporting debris. Observation of major surface flow patterns, by contrast, is best done at the peak of the flood. It is also worth observing the system after the 'worst' of the flooding. The drainage of ponded water often reveals very clearly where there are problems from poor site grading, such as where areas lie below the drain inlets.

Uncovered channels are the easiest ones in which to observe hydraulic performance. A systematic traverse of open channel drainage systems can reveal the effects of bottlenecks such as culverts. Once these are identified, simple tape measurements from the top of the channel to the water level both upstream and downstream of the obstruction can estimate the magnitude of the head loss when combined with ground level data at the points of measurement. While flow measurement during storms is always problematic, engineering judgement is usually adequate in deciding whether to try to ease the constriction.

It is also worth noting how water actually enters the open drain, and whether this can be modified to reduce erosion or the deposit of solids in the drain. One of the advantages of open drains is that if there is a local obstacle, water can enter further downstream.

Closed systems, however, present much greater problems in assessment during wet weather. Inlet blockages can be noted during a systematic traverse of the network. Severe contractions of flow are also an indication that the capacity of the inlet is inadequate. It is also worth noting when water is *leaving* an inlet (i.e. when water is spilling into the road from the inlet), as this is a sure sign that the downstream capacity is exceeded!

During a storm, the flood waves travelling through the system cause water levels to vary rapidly in space and time. This makes it difficult to interpret water level measurements unless they are used as part of a detailed hydraulic study using the dynamic hydraulic models described in Annexe 8-A. In such studies, systematic collection of water levels, with equipment no more sophisticated than a tape measure, can prove invaluable in verifying the quality of the model.

Manholes can be opened for spot observations of water levels, but the cover must be replaced whenever staff are not present. These observations are best made during the earlier and later stages of the storm but, of course, must be abandoned during a flood. The depth from the manhole frame to the level of water can be measured with a tape, and in Indore we were able to do this without entering the manhole. This information can be combined with the frame elevation to compute true water levels. Given the large number of manholes in most systems, it would be unwise to attempt this for every manhole during a storm; it is best to identify a small number of such manholes, and to check them at regular intervals. To be of any use, level readings must be recorded at the correct time, which requires staff to check that all their watches are showing the same time at the start of the storm and to cross-check again at the end of the storm. Waterproof watches are essential!

Surface routes of flow The main task here is to identify the routes on the maps prepared as described in Chapter 5. It can also be worthwhile to estimate the maximum depth of flow and the velocity. This is unlikely to be precise because it will depend on somebody being at the right place at the right time, but it can give you an idea of the relative magnitude of surface flow as opposed to pipe flow. One way to estimate surface velocities is to throw floating objects (such as an orange) into the water, and time the travel between two points a fixed distance apart (Figure 11.1). Dividing the distance by

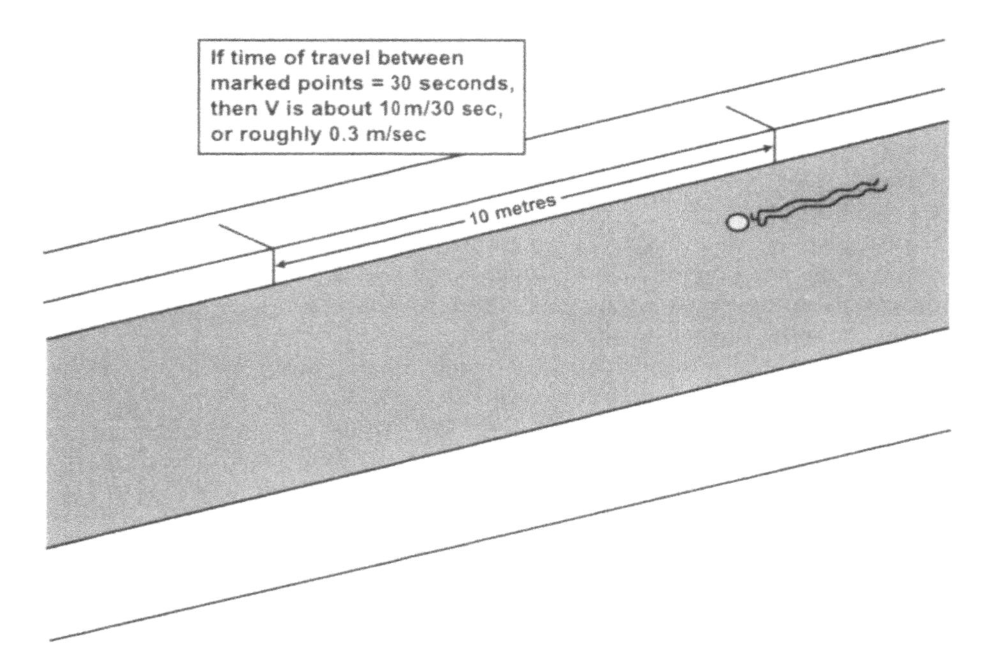

Figure 11.1 *Approximate measurement of surface velocities*

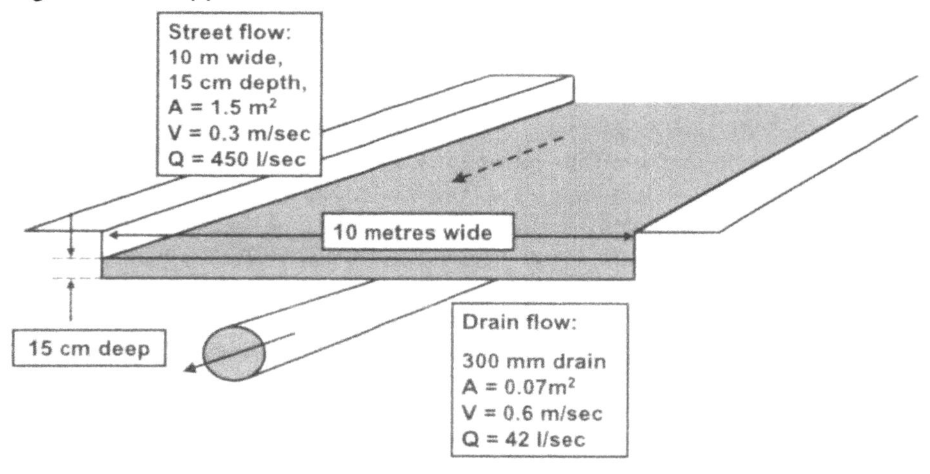

Figure 11.2 *Relative importance of drain and surface flow*

the time yields the velocity of flow. Although velocities will vary across the cross-section, such a measurement permits an estimate of the flow in the street.

Figure 11.2 shows how such estimates can give an idea of the importance of surface flow. In the street shown, with a depth of flooding of only 15 cm, and a velocity of only 0.3 m/sec, the street is carrying over ten times the flow of the drain. Yet 15 cm of 'flooding' above the surface of the road can be tolerated if the water can drain reasonably rapidly, and the thresholds of houses are set high enough above the road surface.

Nuisance, hazard, and damage As noted earlier, community residents are the experts on this question, and storms present a good opportunity to see and hear what they perceive as problems. Structured forms are *not* recommended, without at least some early 'unstructured' observation. Field workers should keep an eye open for the following kinds of problems, and discuss in the debriefing if they've been seen or mentioned by residents:

- flooded homes
- housing damage; nature and extent
- damage to roads (*n.b.* this can be checked *after* the storm)
- backflow from toilets connected to drains
- damaged goods in shops
- damaged furnishings, possessions in homes
- injury (e.g. stepping into open drains, or open manholes)
- lost sleep
- lost work.

Of course, if residents mention other problems, these should be noted; residents may not know much hydrology or hydraulics, but they certainly know what their problems are!

Summary

Table 11.1 illustrates the ways in which different data can be seen or checked at different stages of a storm. In practice, of course, no team can 'count on' a flood occurring, but a team can be ready to look at different aspects if a flood takes place.

Table 11.1 Wet weather observations and timing

	Beginning (or small storms)	*Middle (flood)*	*End (flood water drainage)*
Catchment boundaries	directions of flow in streets	catchment boundaries	catchment boundaries
Flood-prone areas		bottlenecks, causes of flooding	
Hydraulic performance	Inlet performance, blockages, high head losses	outflows from drains	problems of grading, slow drainage, head losses
Surface routes		location, direction, and magnitudes of surface routes	
Nuisance and hazard		Observe, discuss with residents	Observe, discuss with residents

12 Summary and conclusions

Themes of preceding chapters

Why flooding matters

Surface water drainage, also known as storm drainage, removes runoff from the streets and homes of the city. Without good surface water drainage, frequent and prolonged flooding will occur.

Flooding matters for a variety of reasons:

- damage to roads and buildings
- the spread of faecal contamination
- increased risk of accidents (especially for pedestrians who cannot see where they step)
- the creation of mosquito breeding sites
- the economic loss of reduced trade and transport
- the nuisance and mess of clean-up.

Performance

This manual describes drainage system evaluation techniques which are based upon performance, defined in terms of the

- depth
- extent
- duration, and
- frequency of flooding.

The manual addresses two basic questions:

- How much do various factors influence performance?
- How can evaluation of a drainage system provide knowledge that will allow changes in system management to improve performance?

Factors that affect performance

Chapter 3 summarizes available evidence as to the factors that influence performance. The following points emerge as critical:

- *Major and minor drainage systems*. There are always two sets of drainage paths in any given catchment, even if only one of them is designed: (i) the *minor drainage system* of pipes and conduits designed to manage frequent

http://dx.doi.org/10.3362/9781780446059.010

rainfall events, and (ii) the *major drainage system* of surface routes, including roads and open spaces, through which runoff passes when the capacity of the minor system is exceeded.

- *Solids in drains often reduce capacity and performance substantially.* There are many types of solids, including rubbish and building debris, which can end up in the drains of low-income communities in developing countries. These solids are often too large to be 'swept along with the flow', and can lead to a substantial reduction in hydraulic capacity. Under some reasonable assumptions, it can be shown that if the bottom 30 per cent of a drain is filled with solids, its capacity may be exceeded 20 times more often than if it were clean! If there are no major surface routes, so that all runoff must leave the area through a drain, solids blockage can also have a great influence on the duration of flooding.
- *Inlets to closed drainage systems can also be critical to performance.* Ideally, inlets would facilitate the entry of water while obstructing the entry of solids. Inlets frequently perform badly because they are poorly sited (and so do not drain their catchment), or because they are easily clogged or choked. Little attention is paid to inlet design and location, yet systematic evaluation of the network can often point to areas where they are a significant problem.

General approaches to drainage evaluation

There are four principles underlying this manual's approach to evaluation.

- *Evaluations should answer questions*, and the appropriate type and level of evaluation depends upon the questions to be answered, and the resources available to answer them. Given the size and complexity of drainage systems, it is reasonable to invest time in defining questions to be answered, rather than to attempt a 'complete' evaluation that would try to answer all possible questions.
- *Performance matters* and evaluation should reflect those factors in the field most likely to affect the system's performance during wet weather. Field observation during wet weather is invaluable.
- *Measurement of indicators of performance and process is usually more appropriate and useful than measurement of performance itself.* Wet-weather work offers invaluable insights, but the direct measurement of performance (in terms of depth, area, duration, and frequency of flooding) is time-consuming, tedious, and vulnerable to unusual variations in weather. Such efforts are justified in research projects trying to understand the nature of flooding problems, or in large-scale city-wide analyses, but are unlikely to be useful for practical small-scale evaluations. Most evaluations would do better to look at performance indicators (such as the depth of solids in drains, the frequency with which inlets are blocked, etc.) and process indicators (e.g., how quickly are drain cleanings removed? how many staff are available?) as a starting point.

- o Drainage evaluation requires investigation of all the factors that influence performance:
  - the catchment itself
  - the nature of flooding in the catchment
  - the flows that can be expected from rainfall
  - the capacity and condition of the drainage network
  - system maintenance.

Studying the catchment

Defining the catchment area is the start of any evaluation. The definition of a catchment seems like a simple task, but it is often complicated by the absence of good mapping or reliable data on small drains, road and pavement profiles and open ground. Catchment definition is particularly tricky in flat areas, where catchments can actually change during a storm as adjacent areas overflow. Nevertheless, the individual managing the evaluation must establish the limits of the area; the greatest design errors this study encountered arose because of poor definition of the catchment.

Cover surveys are important for estimating the runoff potential of the catchment, as expressed in the C-coefficient of the rational formula. Engineers must also keep an eye on the future; open grassy plots which will be converted to buildings in the near future will increase the runoff in the catchment, and may provoke flooding.

Assessing flooding as a problem

There are a number of ways to do this. Unless detailed hydraulic modelling is proposed, the quickest and easiest way to do this is by asking people in a systematic fashion. Care must be taken that interviewers don't ask leading questions, and that efforts are made to ask more than one person in an area to verify answers about the problem.

For the purposes of hydraulic modelling, electronic depth and velocity gauges are ideal, while chalk gauges are a relatively cheap, simple, and reliable way to obtain maximum water levels above ground.

Estimating flows from runoff

Unless detailed modelling is to be done, the rational method is adequate for estimating flows in peak catchments. Use of the rational method requires knowledge of the type and extent of various types of cover in the catchment (e.g. grass, road, roofing), the catchment area, and the appropriate rainfall intensity to be used.

Rainfall intensity is often the most difficult to estimate. Sometimes fixed criteria (e.g. 25 mm/hr) are used, but in many cases these are arbitrary. Ideally, one can use intensity-duration-frequency (IDF) curves developed from local recording rain gauges. Where these are unavailable, a number of methods exist to estimate IDF curves from daily rainfall data, as described in Annexe 7-A.

Assessing drainage capacity

It is helpful to distinguish between three types of capacity:

- design capacity, based on original design drawings
- as-built capacity, based on the drains as actually built, but with no allowance for solids or blockages
- actual capacity, similar to as-built capacity, but with explicit allowance for solids and blockages, based on field observation.

Design capacity is useful only where good design drawings are available. Actual capacity is less 'precise' than the other two, but is inherently more meaningful, as it reflects the realities of the network's ability to convey runoff.

Drainage network structural survey

Surveys of open systems are relatively straightforward, while surveys of closed conduits are problematic. Powerful lamps and mirrors can help to identify blockages in closed conduit systems, particularly where the distance between manholes is short. The simple presence of standing water above the invert of an outgoing pipe is a sign that drainage is not free and that some obstruction is present.

A good and consistent identification system for manholes and pipes is critical. There may be a temptation to change some numbers or labels as new pipes or manholes are discovered (or some are found to be outside the study area). While new labels have to be added for newly found parts of the system, never change an old label, or use it for a different location; this can only create later confusion in reviewing field notes.

Maintenance survey

Solids levels can be easily assessed in open drains. This is best done by defining the original section at gauging points through measuring the original section, and then measuring down from the top to the solids level.

Measuring solids levels in closed drains is difficult, as solids levels at manholes are not the same as those through the pipe itself. These levels may be estimated qualitatively with the aid of mirrors, but even here much of the pipe is in the dark. Quick qualitative surveys as to the extent of inlet blockage can be helpful to indicate how significant a problem this may be. Drain cleaning piles, where contaminated with sewage, constitute a significant health hazard for children, and these piles should be monitored to see how long it takes for their removal.

Studying drainage systems in action

Field work during and after storms offers insights for evaluation which cannot be obtained any other way. Such work can be most helpful in: defining the catchment itself; identifying the major surface routes of flow, and the direction of flow along them; and the nature and extent of flooding problems. Individual measurements can also be made (e.g. of the variation of water level with time) which are essential to detailed modelling studies.

Such field work must be well planned and organized to make the most of available staff during unscheduled and infrequent storms. Staff need to know the jobs and their locations 'for the next storm' so that valuable time isn't lost trying to find team members and decide the work to be done. It is best to define a sequence of tasks, with the highest priority tasks at the start, so that if rainfall stops quickly the most important work is still done. Different tasks are best done at different stages of the storm, as shown in Table 11.1, and the sequence of jobs can be based on this.

Final conclusions

The ideas described in this manual are based on observation and common sense, and are not particularly complicated. Nevertheless, they conflict with what most engineers are taught about drainage. Conventional drainage design practice assumes that:

- there are no significant solids deposits in drains, as the slope is designed to produce a 'self-cleansing velocity' in the flow
- drains can be designed and built to keep flooding relatively rare, so that design need not address the issue of what happens when it floods.

Field work in Indore, using the techniques described in this manual, established the common sense observations that:

- solids deposits in drains are substantial, and have a major impact on drainage capacity
- the theoretical concept of 'self-cleansing velocities' is completely inapplicable to the size-distribution of solids (including construction debris and solid waste) likely to be found in many urban drains of developing countries
- frequent flooding is inevitable where rainfall intensities are high, the surface does not drain naturally, and solid waste management is poor.

These conclusions point to the frequent overflow of the minor drainage system of pipes and channels into the major drainage system of streets and open spaces.

Implications for improving system performance

The earlier chapters have stressed the need to go out and look at the system as it stands, study blockages of drains and inlets, and observe how the system performs in wet weather. These ideas, combined with the above conclusions, have several implications for system management, for example:

- The quickest and cheapest way to improve performance of an existing system is to evaluate it, identify its weak spots and rectify them.
- It is important to consider both the major and minor drainage systems when setting priorities for drainage interventions; the biggest problems are where major drainage is poor or nonexistent.

- Many systems have bottlenecks where limited capacity affects the upstream system. This problem may be built into the system (where the conduits are too small or the slopes too flat), or it may be because of inadequate maintenance. Effective evaluations can identify these bottlenecks and show the need for their cleaning and/or replacement *in the context of the whole system.* There is no benefit in 'solving' a problem in one place, only to make it worse in another.
- Cleaning an existing drain is usually cheaper than building a new one; why build a new one if the existing one can be cleaned out regularly?
- Careful identification of the processes by which solids enter drains or block inlets may point the way to more effective solids control. Better solid waste management, and better construction site management, may be the most cost-effective ways to improve drainage performance.

Implications for improving drainage design

These observations have more profound implications for drainage design. The most important principle is to consider what surface routes are available and how they will perform. Deliberately *designing* roads to act as surface water drains, as was done in parts of Indore, offers several advantages:

- It provides natural surface water drainage for the housing between roads.
- It is easier to sweep a road than to clean a drain.
- It can eliminate the need for underground storm drains for substantial areas.
- Even where underground drains are built, the roads can act as an effective major drainage system for managing the overflow from heavy storms.

Such an approach cannot be adopted everywhere; topography may dictate the need for pipes and conduits at key locations. The road-as-drain approach can be used only for surface water; sullage, sewage and human excreta must still be managed by other means. It can only be used with confidence on paved roads (although the use of conventional drainage on unpaved areas is also problematic as the soil erodes and chokes the drain). Despite these limitations, the use of roads as drains deserves serious attention when developing new areas, or upgrading existing ones. At the very least, careful site grading and the use of surface routing can reduce the length of drain that needs to be maintained. Further work needs to be done to develop appropriate guidelines for the design of 'road as drain' systems in developing countries, and to assess the current experience of sites where this has been done. Argue (1986) has prepared an excellent major/minor drainage design manual for Australian practice, but appropriate forms of this approach need to be developed for low-income communities in developing countries.

References

Argue, J.R. (1986). *Storm Drainage Design in Small Urban Catchments: a handbook for Australian practice.* Special Report SR 34. Australian Road Research Board: Vermont South, Victoria.

Bartlett, R.E. (1976). *Surface Water Sewerage.* Applied Science Publishers: London.

Binnie & Partners and Hydraulics Research (1987). *Sediment Movement in Combined Sewerage and Storm-Water Drainage Systems*, Project Report No. 1. CIRIA: London.

Breed C.B. and G.L. Hosmer (1977). *The Principles and Practices of Surveying. Volume I: Elementary Surveying.* (Eleventh edition, revised by W. Faig and B.A. Barry). John Wiley & Sons: New York.

Butler, D. and S.H.P.G. Karunaratne (1995). 'Suspended solids trap efficiency of the roadside gully pot', *Water Research*, 29, No. 2, pp. 719–29.

Cairncross, S. and E.A.R. Ouano (1991). *Surface Water Drainage for Low-Income Communities.* World Health Organization: Geneva.

Chow, V.T. (1959). *Open-Channel Hydraulics.* McGraw-Hill: New York.

Chow, V.T., D.R. Maidment and L.W. Mays (1987). *Applied Hydrology.* McGraw-Hill: New York.

Clark, J.W., W. Viessman and M.J. Hammer (1977). *Water Supply and Pollution Control.* (Third edition). Harper & Row: New York.

DoE/NWC (1983). *The Wallingford Procedure: design and analysis of urban storm drainage. (Volume 1: Principles, Methods and Practice).* Department of the Environment/National Water Council Standing Technical Committee Report No. 28. Hydraulics Research, Wallingford.

Escritt, L.B. (1972). *Public Health Engineering Practice.* (Fourth edition). *Volume II: Sewerage and Sewage Disposal.* Macdonald & Evans: Estover.

Fair, G.M., J.C. Geyer and D.A. Okun (1966). *Water and Wastewater Engineering: Volume 1, Water Supply and Wastewater Removal.* John Wiley & Sons: New York.

Feuerstein, M.T. (1986). *Partners in Evaluation: evaluating development and community programmes with participants.* Macmillan: London.

Folland, C.K. and M.G. Colgate (1979). 'Recent and planned rainfall studies in the meteorological office with an application to urban drainage design', pp. 51–70 in *Urban Storm Drainage: Proceedings of the International Conference held at the University of Southampton, April 1978*, ed. by P.R. Helliwell. Pentech Press: London.

Francis, J.R.D. (1975). *Fluid Mechanics for Engineering Students.* (Fourth edition). Edward Arnold: London.

Garg, S.K. (1994). *Sewage Disposal and Air Pollution Engineering.* (Ninth revised edition). Khanna Publishers: New Delhi.

http://dx.doi.org/10.3362/9781780446059.011

Goethert, R. and N. Hamdi (1988). *Making Microplans: a community-based process in programming and development*. Intermediate Technology Publications: London.
Henderson, F.M. (1966). *Open Channel Flow*. Macmillan: New York.
Heywood, G.M. (1994). 'Performance Evaluation of Surface Water Drainage for Low Income Communities'. MSc dissertation, Imperial College of Science, Technology and Medicine: London.
Heywood, G.M., P.J. Kolsky and D. Butler (1997). 'Modelling drainage performance in an Indian catchment'. *Journal of the Chartered Institution of Water and Environmental Management, 11*(1), pp. 31–8.
Huber, W.C. and R.E. Dickinson (1988). *Storm Water Management Model – SWMM, Version 4 User's Manual*. US Environmental Protection Agency: Athens, Georgia (USA).
Hudson, N. (1981). *Soil Conservation*. B.T. Batsford: London.
Khanna, P.N. (1992) Chapter 16, 'Drainage and Sewerage', *Indian Practical Civil Engineers' Handbook*, (13th edition). Engineers Publishers: New Delhi.
Kibler, D.F. (1982). *Urban Stormwater Hydrology*. Water Resources Monograph 7 of the American Geophysical Union. American Geophysical Union: Washington, DC.
Kolsky, P.J., J.N. Parkinson, and D. Butler (1996). 'Third World Surface Water Drainage: The Effect of Solids on Performance', Chapter 14 in *Low-Cost Sewerage* (ed. by D.D. Mara). John Wiley & Sons: Chichester.
Kothyari, U.C. and R.J. Garde (1992). 'Rainfall intensity-duration-frequency formula for India'. *J. Hydr. Engrg. ASCE 118*(2), pp. 323–6.
Malcolm, J. (1966). *Elementary Surveying* (Third edition, revised by D.H. Fryer). University Tutorial Press: London.
Metcalf and Eddy and Engineering Science, Inc. (1966). 'WHO Master-Plan Report' on water supply, sewerage and drainage for the Greater Calcutta area, cited in *Drainage of the Calcutta Area: An Overview* by S.K.D. Gupta in *Calcutta's Urban Future* (ed. by B. Das Gupta *et al.*, 1991). Government of West Bengal: Calcutta.
Moraes, L.R.S. (1996). 'Health Impacts of Drainage and Sewerage in Poor Urban Areas in Salvador, Brazil'. PhD Thesis. London School of Hygiene & Tropical Medicine: London.
Parikh, H. (1990). *Aranya – An Approach to Settlement Design; Planning and Design of Low-Cost Housing Project at Indore, India*. HUDCO: New Delhi.
SCS (1975). *Urban Hydrology for Small Watersheds*. Technical Release No. 55. US Soil Conservation Service: Washington DC.
Shaw, E.M. (1994). *Hydrology in Practice* (Third edition). Chapman & Hall: London.
Sieker, F. (1979). 'Investigation of the accuracy of the postulate "total rainfall frequency equals flood peak frequency",' pp. 31–7 in *Urban Storm Drainage: Proceedings of the International Conference held at the University of Southampton, April 1978*, ed. by P.R. Helliwell. Pentech Press: London.
Smith, C.M.G. (1993). 'The effect of the introduction of piped sewerage on Ascaris infection and environmental contamination in a Gaza Strip refugee camp'. PhD thesis at the London School of Hygiene and Tropical Medicine, University of London.
Stephens, C., R. Pathnaik and S. Lewin (1994). *This is my Beautiful Home: risk perceptions towards flooding and environment in low income communities*. London School of Hygiene and Tropical Medicine: London.

Stern, Peter and others (1983). *Field Engineering: a guide to construction and development work in rural areas.* Intermediate Technology Publications: London.
Tayler, K. and A. Cotton (1993). *Urban Upgrading: Options and Procedures for Pakistan.* WEDC: Loughborough.
Verbanck, M. (ed.) (1992). 'Origin, Occurrence and Behaviour of Sediments in Sewer Systems' (edited special issue), *Water Science and Technology 25*(8).
Verbanck, M.A., R.M. Ashley and A. Bashoc (1994). International workshop on origin, occurrence and behavior of sediments in sewer systems – summary of conclusions. *Water Research 28*: 187–94.
Watkins, L.H. and D. Fiddes (1984). *Highway and Urban Hydrology in the Tropics.* Pentech Press: London.
Watson-Hawksley International, Ltd. and Associated Industrial Consultants (India) Pvt Ltd (1993). *Master Planning for Greater Bombay Storm Drainage & Sewer Rehabilitation (Brimstowad Project), Final Report.* Details available from Montgomery-Watson International: High Wycombe, UK.
WEF/ASCE (1992). *Design and Construction of Urban Stormwater Management Systems.* American Society of Civil Engineers/Water and Environment Federation: New York and Alexandria, VA.

www.ingramcontent.com/pod-product-compliance
Lightning Source LLC
Jackson TN
JSHW011351130125
77033JS00015B/557